2004/2009

理查德·迈耶

Richard Meier Architect

（第五卷）

（美）理查德·迈耶 编
胡一可 译

江苏凤凰科学技术出版社

图书在版编目（CIP）数据

理查德·迈耶. 第5卷 /（美）迈耶编；胡一可译.
-- 南京：江苏凤凰科学技术出版社，2015.8
 ISBN 978-7-5537-5252-5

Ⅰ.①理… Ⅱ.①迈… ②胡… Ⅲ.①建筑设计—作品集—美国—现代 Ⅳ.①TU206

中国版本图书馆CIP数据核字(2015)第190789号

江苏省版权局著作权合同登记图字：10-2015-268

Copyright © 2009

Originally published in English under the title Richard Meier Architect, Volume 5 in 2009. This translation published by agreement with Rizzoli International Publications, New York through the Chinese Connection Agency, a division of The Yao Enterprises, LLC.

理查德·迈耶（第五卷）

编　　　者	（美）理查德·迈耶
译　　　者	胡一可
项 目 策 划	凤凰空间/陈　景
责 任 编 辑	刘屹立
特 约 编 辑	赵　萌
出 版 发 行	凤凰出版传媒股份有限公司
	江苏凤凰科学技术出版社
出版社地址	南京市湖南路1号A楼，邮编：210009
出版社网址	http://www.pspress.cn
总 经 销	天津凤凰空间文化传媒有限公司
总经销网址	http://www.ifengspace.cn
经　　　销	全国新华书店
印　　　刷	北京博海升彩色印刷有限公司
开　　　本	636 mm×939 mm　1/16
印　　　张	26.5
字　　　数	212 000
版　　　次	2015年8月第1版
印　　　次	2015年8月第1次印刷
标 准 书 号	ISBN 978-7-5537-5252-5
定　　　价	298.00元

图书如有印装质量问题，可随时向销售部调换（电话：022-87893668）。

目录 Contents

序
Preface by Richard Meier 4

风格是一个没有复数形式的词
Style Is a Word That Has No Plural 6

私人建筑及工程项目
Private Buildings and Projects

乔伊公寓
Joy Apartment 14

马利布海滩住宅
Malibu Beach House 22

查尔斯街165号
165 Charles Street 34

展望公园边的建筑
On Prospect Park 54

南佛罗里达住宅
Southern Florida House 62

深圳住宅
Houses in Shenzhen 70

瑞克麦斯住宅
Rickmers House 88

第五大道公寓
Fifth Avenue Apartment 94

卢森堡住宅
Luxembourg House 102

博得鲁姆住宅
Bodrum Houses 110

2号住宅
House 2 116

天津别墅
Tianjin Villa 126

公共建筑与项目
Public Buildings and Projects

阿尔普博物馆
Arp Museum 138

和平祭坛博物馆
Ara Pacis Museum 158

圣何塞市政厅
San Jose City Hall 174

加州大学洛杉矶分校艾利与艾迪斯·布劳德艺术中心
Eli & Edythe Broad Art Center, UCLA 186

布尔达收藏博物馆
Burda Collection Museum 194

皮克-克洛彭堡百货公司
Peek & Cloppenburg Department Store 214

康奈尔大学威尔大厅
Weill Hall, Cornell University 224

ECM城市大厦
ECM City Tower 242

Feldmühleplatz办公楼
Feldmühleplatz Office Building 254

耶索洛丽都村、公寓房和酒店
Jesolo Lido Village, Condominium, and Hotel 262

圣丹尼斯办公楼
Saint-Denis Office Development 280

美国联邦法院
United States Courthouse 288

东江总体规划
East River Master Plan 298

咖啡广场
Coffee Plaza 312

9900威尔希尔
9900 Wilshire 320

意大利水泥创新技术中心实验室
Italcementi Innovation and Technology Central Laboratory 330

福斯特曼·里特尔公司和国际管理集团世界总部
Forstmann Little and Co. and IMG Worldwide Headquarters 338

罗斯柴尔德大厦
Rothschild Tower 346

纽瓦克苏玛总体规划
SoMa Newark Master Plan 356

天津酒店
Tianjin Hotel 364

贝多芬音乐节剧院
Beethoven Festspielhaus 372

编后记
Afterword by Paul Goldberger 384

附言
Postscript by Frank Stella 387

理查德·迈耶简介
Richard Meier 389

年代表
Chronology 390

理查德·迈耶及其合伙人事务所模型博物馆
Richard Meier & Partners Model Museum 408

参考文献
Bibliography 416

合作者
Collaborators 422

顾问
Consultants 423

图片注解
Illustration Credits 423

序
Preface

理查德·迈耶

我很小的时候常常在周六去探望我的祖母,她的公寓在新泽西州东奥兰治市。她会用一些简单的材料制作美味的薄饼给我吃,享受这样的美食给了我难以形容的快乐。阳光洒下来,我们坐在祖母的厨房里,她会向我讲述她96年生命里所遇见的所有奇迹,描述她见过的那些革命性的发明:电灯、电话、电视、飞机。她的故事让我震惊。我不相信一个人能经历这么多。当然现在我能更容易地理解这些事情了。

现在我回过头来看,这些思考从我写第一卷引言就开始了。当我们开始汇总第一本书的所有内容时,我们的实践团队刚初具规模,只做了一些小型实践项目。我很兴奋我们有足以支撑一本书的材料,但从未想到此书能出到第五卷。很多事情都改变了,如今很少有建筑师画图了,偶尔动笔也是做数据录入而非绘制草图。用电脑,你可以瞬间做好图纸的更改、重绘。以前承包商们无法使用手绘图纸,现在的电脑图纸让他们可以方便地使用。但是建筑设计仍然是人的创造力的产物,人的大脑通过构想来创造供人停留的和穿越的空间,需要依靠灵感和思想完成,从事建筑设计项目的建筑师或者设计团队必须深谙此道。

建筑学的特点未曾改变,但是世界上的很多其他事物已然转变,对当今抑或对未来的预测将变得愈发困难。建筑面临诸多挑战,这些挑战以对环境负责的态度呈现,与公共领域的实体空间形态产生关联。建筑必然是开放的体系,能够缓解城市未来面临的巨大压力。建筑还必须处理城市设计问题,促进社会关系发展,协调公共空间与私密空间的关系。

虽然在大多数的项目中建筑师会与客户合作,但他们才是创造那些长久存在事物的真正的主导者,建筑师促成的是比所建造实体自身更加伟大的事物,他们促成的是比任何参与创作的工作人员的寿命更加长久的事物。要产生建筑创作灵感,建筑师不仅要思考建筑文脉、场地环境、历史、周边建筑、地形、公共领域的性质,也要思考建筑能是什么、将是什么,以及它会对未来社会产生怎样的意义。

伴随着当今世界上发生的一切变化,建筑持续地以其美学特质感动着我们,就如同过去那些伟大的建筑一样,这一点从未改变。我期望我们能重新审视这个世界,探索新的可能,这样我们可以改变思考和居住的方式,不论针对个人还是群体。

我要介绍我在洛杉矶工作室的合作伙伴迈克尔·帕拉迪诺;纽约的合作伙伴伯恩哈德·卡普弗、伦尼·洛根和严德浩;洛杉矶的吉姆·克劳福德和里克·欧文;还有为此卷中所有项目倾注心血的纽约和洛杉矶工作室的全体成员。

虽然此书名为《理查德·迈耶(第五卷)》,但是这些年同事们相聚并组成极为优秀的团队是一种缘分,没有他们的帮助,这项工作根本无法完成。我们不仅有幸拥有如此杰出的项目组成员,还有在纽约和洛杉矶工作室一起工作的那些令人惊叹、独具创造力、热爱挑战的顾问团队。

这些项目的完成也依赖业主的宽容与信赖。我们非常幸运能与这样的客户合作,每次的努力都能获得新的合作伙伴。多年来,更有一位孜孜不倦的建筑评论家和一位历史评论家给予工作室项目巨大的帮助,我要再次感谢保罗·戈登伯格和肯尼斯·弗兰普顿具有思想性和洞察力的文章。

第三位我要介绍的建筑评论家也是一位艺术家,他为此卷撰文,更重要的是,他是我五十多年的挚友,他与我分享的那些经历激发了我的创作灵感。弗兰克·斯特拉的建筑实验,以一种初始的、革新的、煽动性的方式影响着建筑未来发展的进程。

我衷心感谢他,我们这个时代最好的平面设计师之一,马西诺·维格纳里超凡的天赋和敏锐的视觉设计,此卷中你看到的所有成果就不会如此优秀。与马西诺一起共事总是令人愉悦,同样,比阿特里斯·西风特斯负责每一页细节的排版设计。我很感谢他的辛勤奉献。

整合图片、图纸和文字需要智慧和热情,基于此我要为玛丽·露·邦恩的才华喝彩。当然,这里还有很多在审视建筑、捕捉空间光影方面独具慧眼的摄影师,他们的工作异常重要。斯科特·弗朗西斯、罗兰·哈博、乔克·波特儿是与我们长期合作的三位艺术家,他们的摄影作品捕捉到了我们多年来一直努力达到的设计效果,我非常感谢他们,也非常感谢为此卷作品拍摄的其他摄影师们。最后,我要感谢我的女儿安娜,对我来说她意味着一切。

风格是一个没有复数形式的词
Style Is a Word That Has No Plural

肯尼斯·弗兰普顿

风格就像一道彩虹。它是由特定实体引发的知觉现象,我们只能在阳光明媚与阴雨连绵两种天气转换的瞬间捕捉到它,当我们到达我们认为看到彩虹的地方,它也随即消失。每当我们认为抓到了彩虹,如同画家创作的画作一样,它便消失在这位画家的前辈或者追随者的作品中,这种现象会在画家的作品中反复出现,所以这位画家的青年和老年阶段的作品,以及他的老师和学生的作品,都充满了潜在的、如化石般永恒的东西。现在什么是有效的?是以完全实体形式存在的孤立的作品,还是在特定范围内的所有作品?风格属于对实体的静态组合的评价。一旦这些实体随着时间推移重新呈现,风格就消失了。

乔治·库布勒,时间的形状,1962

45年的实践,毫无疑问产生了如此多的建成建筑,更不用提相同数量的未建成方案和总体规划成果,这是一个以任何标准评价都算是多产的职业生涯,也是一个见证作品范围和规模随着时间发生巨大变化的职业生涯。迈耶在壮美的自然场景中设计了明亮华丽的住宅——道格拉斯住宅(1973)——这幢住宅成为了最经典的案例之一,也为他赢得了实至名归的声望,除此之外,20世纪80年代中期,迈耶在美国和欧洲,以公共建筑师的特殊身份展开自己的设计工作。除了带来决定性成就的法兰克福装饰艺术博物馆(1985),以及同年洛杉矶盖蒂中心的初始方案,迈耶作品集第三卷涵盖了1992至1999年的设计作品,将会进一步证明迈耶处于他的巅峰状态,如巴黎Canal+总部(1992)、乌尔姆市政厅(1993),还有海牙市政厅(1995)和巴塞罗那当代艺术博物馆(1995)。但是,没有人能说清在此期间迈耶的个人工作遭遇到何种困境,得克萨斯州达拉斯市拉乔夫斯基住宅(1996)充分证明了这一点。

拉乔夫斯基住宅从一开始就被设想成为一个小博物馆,其在实与虚、图与底、公共与私密方面达到的动态平衡状态是迈耶设计的其他住宅无法比拟的。

在迈耶作品集的第四卷中,两个规模巨大的美国法院项目维护了他作为一名市政建筑师的声誉,项目建成于千禧年之际,分别坐落在北美大陆的两端,一个位于纽约州艾斯利普市,另一个位于亚利桑那州菲尼克斯市。从此开始,迈耶建筑的整体形象开始改变。设计室开始从新纯粹主义分离出来,接受了一种更加非物质化的表达方式,使用轻质金属百叶窗,以满足更高的环境控制要求。但是,迈耶的佩里街公寓楼,幕墙表面,共16层,这幢建筑并未使用百叶窗来维持水晶般的效果。位于曼哈顿西村边界哈得孙河边的这两幢双子塔将会被第三栋建筑连接,那是在视觉上使用相同思路却更加精致的16层的查尔斯街公寓楼。最终的结果就是三栋优雅的建筑并置,每栋建筑都有一个立面面向河流,三个立面展现了相似的美学特征——条纹式的肌理和水晶般的效果。该特征迥异于以往40余年纽约开发商使用的沉闷的高层住宅模式。虽然可能有人会批评建筑师没有很好地解决光线调控问题,但建筑采用中空玻璃并且设置可手动操作的开启扇,建筑的空间处理仍然很优雅,幕墙划分的比例仍然很完美。回想起来,这三座高层已经被视作应对环境的恰当结果,在这个环境中天才般的建筑师、坚定的客户、良好的市场环境一同促成了高品质的建筑,也促成了微型城市形态的典范。

查尔斯街区是罕见的机遇,建筑师有机会使它发展成为一种原型。除了精心的规划、宽敞的公寓,查尔斯街区最有趣的地方就是公共设施集成在建筑的裙房中，包括健身房、环形水池、桑拿房和放映室。我们只希望这个原型可以重现，为目前零散、不成系统的临河界面提供令人耳目一新的一个亮点。从15层公寓楼宽阔的界面俯瞰布鲁克林展望公园，观景效果具有均好性，公寓除了为每户提供阳台和方便的地下停车外，并未考虑查尔斯街的公共需求。走廊空间狭窄，公寓每层的房间众多，方案创作的难度不仅来自狭窄的走廊，还来自于将相当数量的公寓集成在建筑的同一层。14、15层的复式公寓中的屋顶花园弥补了公寓缺乏开敞空间的不足。查尔斯街上钢结构的优雅显而易见，佩里和查尔斯街公寓的垂直语汇在整个建筑中形成分散体系，在镶装玻璃的阳台和在建筑转角及其他突出部位的阳台之间形成不稳定的形式，在这些地方，通过非对称的变化打破体块的韵律感。

近五年工作室研究出的最引人注目的住宅开发成果当数意大利威尼斯附近的耶索洛丽都村，自该项目第一阶段实现以来已经完成了许多后续发展阶段的工作。目前，这个复杂体系包含了三个住宅街区，每个街区都由三层高的住宅构成，东西两侧设置露台，刚好利用街区之间的空间设置公共的日光浴平台和游泳池。轴向布局终止于南边等高的横向建筑体块，被非对称的门廊打断，体块下面形成了街区体系，通往海滩主要道路。

这些三层的居住单元面向中央日光浴平台的面被连接起来，由地面露台住宅和一、二层公寓巧妙地组合，而它们的平面几乎是相同的。唯一的变化就在于这些公寓的整体构成上是拥有一个浴室还是两个浴室。地面露台和抬升的单元在空间上的连接方式让人联想到勒·柯布西耶马赛公寓包含两层空间的单元，但这并非有意为之。马赛公寓空间连接之处在剖面上，是不可见的，耶索洛丽都村项目是在平面上连接且是可以直接感知的。两个案例中，存在风险的是空间上的相互影响，这种影响成为整体而非单元的隐含的象征。联排住宅的类型按照固定的组合进行组织，其中的变化和独具趣味之处就在于房子两边露台的处理方式不同，使其在体验上具有差异。因此从一边可直接进入露台单元内，另外一边通过三层高的露天走廊可进入综合体。虽然通道被赋予第二层意义，但是它的主要功能是给公寓一、二层提供楼梯。布局需要一、二层的阳台在同一水平面上并跨越一层露台。在此位置上，它们被水平百叶窗遮挡，不仅为了控制光照，也为了在水平方向上保护各房间的隐私。

缩短公共日光浴平台轴线的南北向体块，使用了完全不同的空间结构。三个连续的楼板上，每个单元均包括卧室和私人阳台，每个单元通过曲折的楼梯成对相连。在此情况下，日光浴平台上的起居厅均为南向，阳台被百叶窗很好地遮挡。

这个正在进行的最新扩建项目是一个10层的"交叉天桥"，双面临街。东西向的公寓建筑体块有效地取代了原有总体规划的25层高层公寓。位置靠近海滩，这栋建筑的标准层包括7个东西向双居室，两个面朝南向海面的三居室。所有的公寓通过曲折的楼梯和一个小客梯成对连接。除了南向的高级单元，相邻公寓的起居室交替地面向东面和西面，从而确保每个公寓的私密性。

同时还有保护私密性的其他手段，例如地面单元和顶层的两层单元，所有的起居室都面向西面，或者在顶层的复式单元中，东西向通透。该街区布局紧凑，在一层配备了公共设施，包括酒吧、游泳池/阳光浴平台，这个空间作为建筑到海滩连续空间的延伸。

最初的整体规划是用一个高层组群作为南北轴线的终点，高层组群由一栋公寓塔楼和俯瞰亚德里亚海的酒店组成。最终这个综合体能否以最初整体规划所预期的那样有条理地运行，尚未可知。

在过去的几年里，工作室收到了两笔单独的佣金，客户委托我们在高度起伏、多山的场地中设计殖民时期风格的别墅。早在2005年，第一个别墅项目是在中国深圳的一个小山顶上设计5栋别墅。2007年，第二个别墅项目是在土耳其博得鲁姆半岛上，在雅乐卡瓦克村外，设计21栋别墅。在深圳的大型奢华别墅项目的设计周期为一年，面积从600平方米到1000平方米不等，雅乐卡瓦克项目面积为中国深圳项目的二分之一到三分之一，主要作为度假别墅使用。不论两个项目之间名声或者总体规模的差异如何，它们都是使用类型学方法进行设计的范例，让新纯粹主义住宅更加完善，因为这是由迈耶1967年史密斯住宅原型和1973年道格拉斯住宅原型发展而来的。这种原型的主要属性包括：第一，两层通高的起居室，连带夹层和玻璃墙前面独立的圆形柱体；第二，服务体块覆盖次要空间，例如卧室、厨房、浴室等，这些空间被带有窗子的实体墙围合。

在深圳，所有的住宅都面向东南，出挑很大的悬臂结构为主要空间遮挡阳光和雨水，别墅通过车库的前院和大的前厅从北面进入，通过楼梯通往两层高的起居室，前院和前厅的一边是水池和平台。在博得鲁姆，相似的模式在延续，在那里有五种不同的原型。本案中，虽然所有的别墅都有日光浴平台和游泳池，但是设计过程迥异，因为设计需要在高低起伏明显的场地中以如此贴近的距离布局建筑。虽然五个原型实际上是标准的，并被多次重复使用，但在某些情况下，从二层进入别墅是必要的，以便适应陡坡环境。在平面中，本案与深圳项目建筑入口均通过相似的几何变形手法处理，保持了从车库进入别墅的原则。除了深圳项目必须有一个直升机停机坪的特殊要求以外，深圳项目与博得鲁姆项目最主要的区别就是在场地中的种植设计，其目的是为了避免别墅之间的视线干扰。关于这些住宅项目最值得一提的就是设计了一个花园城市理念的现代版本，就像现代建筑的那些范例，如同20世纪20年代，1926年建在德绍，包豪斯的创始人沃尔特·格罗皮乌斯的住宅。

经典的迈耶别墅伫立在场地上，当它遇到致密的小镇建筑环境时就会以有趣的方式弯折，就像四层瑞克麦斯住宅那样，瑞克麦斯住宅在德国汉堡一个狭小的场地上，项目始于2003年。这栋建筑的特殊之处在于空间的流动性——一个标高层到另外一个标高层空间以一种流动的方式展开。楼板由宽敞的楼梯和电梯连接，墙体要素有节奏地调整改变，形成了空间的流动性。这块狭窄场地面临的主要问题是如何寻求一种方式清晰地表达不同品质空间的分布，同时保证主要房间避开东南方房地产项目的干扰。

在这里我们运用了俄罗斯套娃的策略，构建建筑中的建筑，部分透明玻璃开窗用独立的乳白玻璃片遮挡。除了调节阳光渗入量，这些乳白玻璃片提供了视觉私密感同时保证全天候的发光照明。外层膜也制造了棱镜体内外之间微妙的体积游戏，同时，一天的日光变化通过乳白色的墙体塑造了五光十色的外形特征。内部宽大的空间让建筑师可以将自由平面的概念和古典纵向构图结合起来。在这里我们看到了一系列比例优美，自由流动的空间，从临街空地延伸到花园内，这样可以为连续的功能空间提供适宜的环境。因此观者可以从临街一侧符合人体工程学要求的厨房进入正式的用餐空间，随后进入宽敞的起居室，与之衔接的是可以俯瞰湖面的被抬升的狭窄的花园阳台。楼上的卧室层也被赋予了类似的空间流动特征，每间卧室都带有成套的盥洗室。顶部三层全部被豪华主卧套房占据。

迈耶工作室有着超常的能力，以专业水准将北德的贵族式住宅的空间特质嫁接到以色列的39层住宅塔楼中。罗斯切尔德塔楼是一个技艺精湛的案例，很难发现可以和它媲美的构造精密的高层住宅塔楼案例。罗斯切尔德塔楼位于特拉维夫，在2010年完工。

在这里，迈耶套用的设计与查尔斯街区的优雅相匹配，四面仅仅恰当地使用百叶遮阳窗作为立面。这些幕布围合成棋盘式的平面，在塔楼开放的角落附近相互铰接。公寓的阳台巧妙地运用这些角落上的屏障进行保护，在剖面上从角落的一侧衔接另一侧。除了占据整层的豪华公寓和建筑顶层的复式公寓，每层都有四个三居室或者两个四居室单元交替分布，从底层到顶层住宅类型以波动的韵律进行分布。

罗斯柴尔德塔楼的设计目标是要成为一个突破性的建筑，不仅仅是为了迈耶的工作室和特拉维夫这个城市，也是为了在全球范围内展现商业地产高层住宅的通用模型。就这一点而言，这一目标似乎并没有实现。项目在规模上进行了调整：一层台基调高，面临爱伦比街的一面削减5层。最贴近这个项目形式的建筑无疑是伦佐·皮亚诺于2006年为纽约时报设计的位于曼哈顿的60层办公楼。这两座建筑之间的相似性也体现在调整百叶窗之间水平间隔这一常见策略上，目的是为了提供全景视野，同时为建筑遮蔽阳光。

罗斯柴尔德塔楼概念上的活跃还体现在它所处的位置与一个精致的8层办公建筑平行，这个办公建筑现在计划成为捷克斯洛伐克的布拉格市中心的核心建筑，该建筑三个暴露在阳光下的建筑立面将要安装百叶窗，北面仅以幕墙覆盖。

项目纲领性和符号性的内容就是构建一个顶层明亮的中庭，这一想法是从弗兰克·劳埃德·赖特1904年的拉金大厦到诺曼·福斯特1975年在伊普斯维奇设计的威利斯·法波尔和大仲马总部或者1997年在法兰克福设计的德国商业银行借鉴并长期演化而来的。这一发展进程的推动力是用中心统一的空间来整合开放式的、有一定等级体系的空间，中心空间对社区和市民活动具有重要意义。设计者用位于中心、顶层明亮的自助餐厅和从入口轴线拔地而起直至顶端的三个符号式的圆柱体将设计意图贯穿在方案中。

水平百叶窗或贯通的平面，或作为结构表皮上大面积的条纹，有助于加强迈耶现阶段设计实践的亲民特征。在2006年完工的罗马和平祭坛博物馆和2004年完工的德国巴登巴登市布尔达收藏美术馆这两个纪念性建筑中，这一特征也十分明显。罗马和平祭坛博物馆实际上是将原来公元9世纪的祭祀空间转化为一个新的城市环境，祭坛直至今日都被闲置在台伯河岸边。德国巴登巴登市布达收藏博物馆是对现存新古典主义博物馆的加建，旨在提供一个独立的艺术品收藏处，地点在巴登巴登市乡村风格的里奇腾塔勒大道上。两个方案的任务都是在敏感的历史文脉中植入现代设计，然而设计策略并不相同。布达拉收藏美术馆保留作为个人馈赠物的辨识度，与此同时，也作为原先存在的美术馆的一部分，用这样的方式为大量的藏品提供空间；而和平祭坛博物馆，需要台基和围护亭，以为古老的历史遗迹提供庇护，引导参观者进入并为这件巨大又神秘的物品而沉思。

在巴登巴登市，在文脉方面的挑战是如何与现存美术馆的新古典主义石材表面呼应而又不被它完全主导。设计应对了这项挑战，大部分新建筑以石材为饰面，同时把它当作针轮平面的一种离心扩散来处理，当观者在两栋建筑之间穿行时，可以体验这种针轮平面产生的恰当、宜人的空间序列。为了这一目的，设计使用轻钢桥将新建筑北侧与老美术馆相连，与主入口的夹层体块结合。同时，曲折的斜坡将轻钢桥形成的轴线进一步延伸，激活了整个入口序列，从同一标高的最初入口到主楼的附加体块构成了入口序列。除了灯槽和建筑内部空间位移的部分，主要的建筑语言是玻璃幕墙外的百叶窗，它们覆盖了主要展览空间的南立面。在这里，就像在瑞克麦斯住宅内部，我们再次看到了建筑内的建筑。

和最初的设想相比，罗马和平祭坛博物馆和它的场地有着更多起承转合的关系，因为，神秘的圆形奥古斯都墓地有着错综复杂的边界，周围环绕着柏树，博物馆正是处于这样的环境之中。当观者离开这个装饰繁复的空间，穿过一个厚重的长方形会堂（对博物馆入口具有屏蔽作用），他不仅会发现一个纪念性的斜角楼梯可通往建筑的台基，还会发现台伯河岸边路堤和奥古斯都大帝广场的标高存在巨大的差异，剖面上的变化让底层的管理处和电子图书馆能够获得自然采光。

显然，成排精致的图案丰富了祭坛的表面，除此之外，设计还有三个值得我们关注的点。首先，面对台基的现存石灰石墙面是受墨索里尼指示，于1938年按照奥古斯生平事迹雕刻的，颇具古典韵味。其次，我们关注建筑主要体块本身的结构，当然，这是工作室完成的最精确的建构表达：4×10肋形楼板梁网格呈现出横向上ABBA的韵律，位于四个柱子以上，端部铰接以承受双向结构的主梁。最后，行进序列的终点是一个两层共140座位的立方体讲堂。顶部是一个向北面倾斜的锥体光井，这显然是迄今为止迈耶完成的最温馨和比例最精致的市民空间之一。虽然迈耶在设计该案室内空间的时候，采用类似阿尔托设计的木条观众席，但是和平祭坛博物馆的屋顶建构让人回想起马拉苏提和曼贾罗蒂设计的巴拉赞提教堂，该教堂于1958年完工，坐落在米兰附近。迈耶是混凝土结构的建构权威，闪耀的玻璃幕墙对称地设置在混凝土结构两边，贯穿整个高度的层层乳白色的水平百叶可遮挡阳光，也激活了混凝土结构，百叶让和平祭坛博物馆成为了一个明亮的体块，但光照强度被严格控制。

直至今日，迈耶的实践在不断地转变，不仅仅在尺度和主题方面，也在光线的调节方面，他的实践也在持续地带来一种语法式的表达。这一整体的改变可能在高层结构中最为明显，迈耶被邀请设计此类结构的频率越来越高。这种改变也可以在近期关于豪华低层住宅区地形的文字描述中看出来，如近期的深圳住宅和博得鲁姆项目。工作室的所有成员不断提升其专业能力，以前，团队的创作能力大多集中在私人住宅和公共博物馆设计方面，现在需要处理更多类型的人居环境肌理。正如我们所见，这种在项目操作中的改变似乎都伴随着一种在"建筑风格"方面的自然而然的变化，因为工作室已经把之前对新纯粹主义的偏好远远抛在身后，去拥抱更加明亮的格调，这种格调中，空间规划的合理性和材料的开发与一种微妙的表皮表达方式相互补充，而且这一方式作为原型是可持续运用的。因此，玻璃幕墙的使用频率越来越高：幕墙由多层乳白色的玻璃或独立的铝百叶窗进行调节。这样做，迈耶的发展路线并未失去它可塑的潜力，迈耶就如同之前一样用柯布西耶留下的财富激发自己的灵感。而现在这个参照物被一种表皮建构形式所取代，这种表皮已经成为迈耶项目的特色，同时也可以改善建筑的环保性能。

私人建筑及工程项目
Private Buildings and Projects

乔伊公寓
Joy Apartment

乔伊公寓
Joy Apartment

纽约市佩里街176号，2001—2006

这栋豪华公寓占地约975平方米，共三层，位于佩里街住宅小区南侧。起居空间，包括两层通高的结构，占据了整个中间层，上层是主卧套房，下层是儿童卧室。儿童卧室层还设有南向的藏书室，兼作会客室套间；顶层为健身室和学习区，可俯瞰起居空间。每一层的中心都有一个服务核，提供浴室、更衣室、衣橱、桑拿和开放的洗浴区。内部旋转楼梯连接三层空间，此外还有电梯可到达每一层。

主要的起居空间被分成一系列活动区域，围绕着中心服务核连续布置。这些空间沿顺时针方向依次为：客厅阳台、旋转楼梯、早餐区、厨房、视听休闲区，最后是起居室。通高的玻璃分隔板贯穿始终；（迈耶）设计了一个非常特别的可滑动的屏风将会客区与藏书室完全分隔。

住宅中的每一个房间都有一个AMX（安玛思）界面的触摸屏，它能管理室内环境，包括光照、遮光窗帘、温度、湿度和音乐播放。高清等离子电视机可以通过电子操控的方式从卧室和起居室之间的栏杆升起。这些设计中的高科技手段对客户来说是关键的因素，（我们的）客户被认为是现代科技的领袖。

下层平面

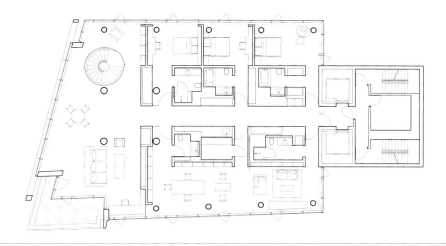

中间层平面

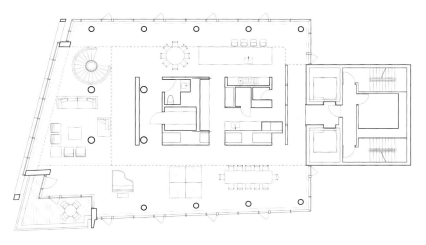

上层平面

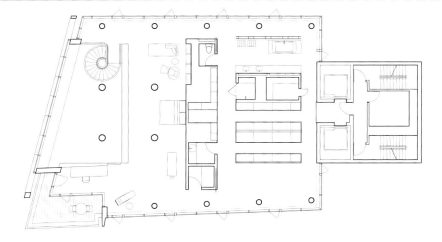

马利布海滩住宅
Malibu Beach House

马利布海滩住宅
Malibu Beach House

加利福尼亚州马利布，2002—2006

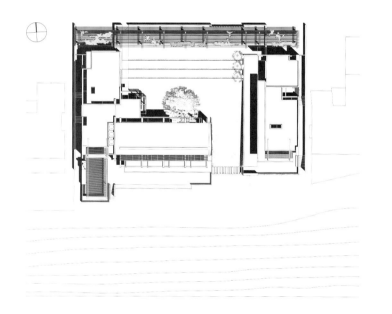

毫不夸张地讲，设计海滨住宅是一种荣幸。这栋住宅及客房项目，和朝南的马利布海滩的其他房产对比，在尺度上和临海面积上是特殊的。

场地设计旨在将周边区域的潜力最大化，这种潜力是十分珍贵的。设计将建筑分成各自独立的结构，海滩的沙子和草地被移植到入口的院子里，成为外部景观的延伸。两栋建筑都能看到海和庭院，被一层可操作的百叶窗滤景和框景，百叶窗独立于建筑围护结构之外。

房子表皮由柚木构成，柚木也是室内地板和顶棚的材质。现场浇筑的混凝土墙将平面一分为二，它也是从街道标高的入口大门到海洋大门和通向海滩的楼梯之间的实体连接。精细的青铜顶、精致的金属构件和各色的材料（混凝土、柚木和青铜）可抗风化、抗氧化，以应对海滨严酷的环境。

围合

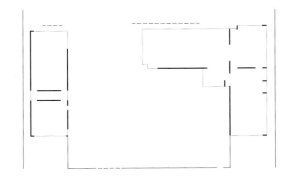

结构

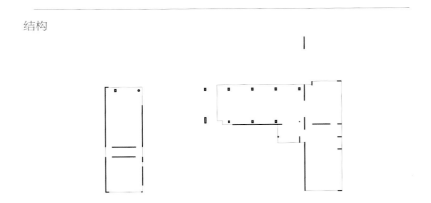

入口

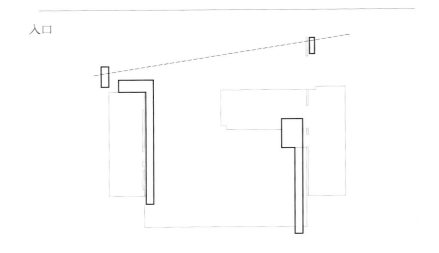

一层平面

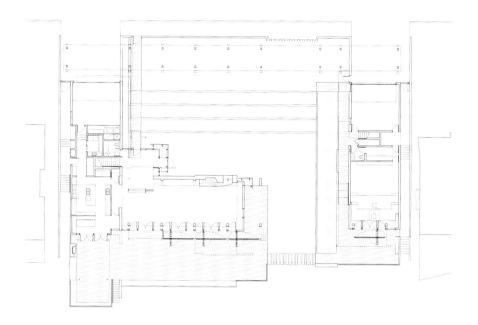

二层平面

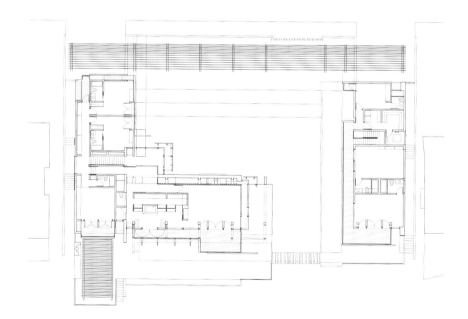

0 10 25

北立面

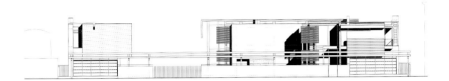

南立面

查尔斯街165号
165 Charles Street

查尔斯街165号
165 Charles Street

纽约州纽约市，2003—2006

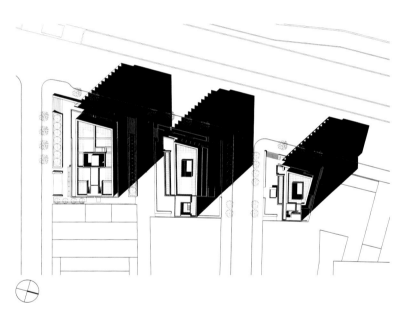

2002年，理查德·迈耶及其合伙人事务所完成了该项目，以延续城市形态和佩里街173/176号建筑那种水晶般的特征。建筑共16层，内含公寓及相关配套设施。除了底部的两层和顶层公寓，这栋建筑的每层都被核心筒分成了一对两居室公寓。与佩里街的开放式阁楼空间相比，这些公寓的空间经过了设计师的精心组织。每间公寓的前面自由地组织一个岛式厨房单元，该单元让起居空间与相邻的学习空间（也可作为餐厅使用）相互渗透，形成流动空间。每两个卧室均配有一个浴室，在沿着核心筒的走廊处开门。

二层包含四套公寓，其中两套有通高的起居室。通高的起居室使三层面积减少，留下了两个一居室的充足空间，这两个一居室，分别为南、北朝向，东面是两套SOHO公寓。

除了常见的水晶般透明的建筑语汇，这个住宅区的立面与佩里街住宅区的造型非常不同，尤其是缺少阳台和交通核的表面处理方法。受查尔斯街相对狭窄场地的影响，设计采用了更加综合的方法。

流线

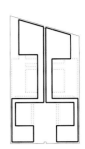

几何分析

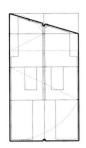

层次

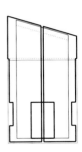

图解

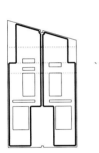

结构

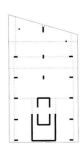

对称

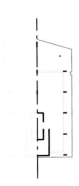

入口层平面

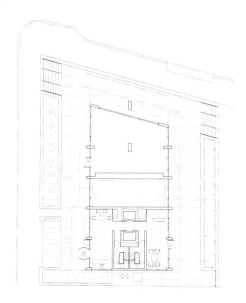

三层平面

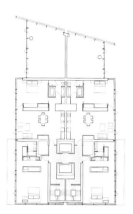

顶层公寓底层平面

顶层公寓上层平面

0　10　20

西立面

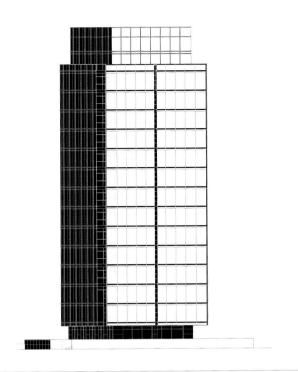

南立面

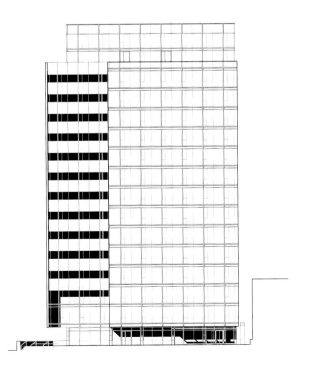

展望公园边的建筑
On Prospect Park

展望公园边的建筑
On Prospect Park

纽约布鲁克林,2003—2009

这个地标式的独立产权公寓位于著名的布鲁克林展望公园的一角,毗邻大军广场,拥有全景式视野,包括展望公园、曼哈顿天际线、纽约港和自由女神像。附近的景点和历史遗迹包括布鲁克林公共图书馆和布鲁克林植物园,当然还有约213万平方米的展望公园。展望公园是由弗雷德里克·劳·奥姆斯特德和卡尔弗特·沃克斯在完成纽约中心公园之后不久设计的。这个场地三面被绿树成荫的宽阔街道环绕。东面,现存住宅的褐色砂石墙和被称为联合神殿的宗教机构,共用建筑红线。北面的圣约翰宫殿与建于20世纪初的4层住宅毗邻。

15层的独立式产权建筑包含约97.5平方米的一居室、约227.6平方米的三居室和复式屋顶公寓。事实上,所有的公寓都有阳台,可欣赏不同方向的壮观景色。地面层为住户提供了约371.6平方米的休闲空间,设有放映室和特殊活动空间。(建筑中)有八个私人屋顶露台,还有一个可供所有住户观赏的屋顶花园。

建筑旨在提升大军广场的城市空间特征,同时建立城市主要干道和历史文脉之间新的关系。本案的基础是尊重现有尺度和周边文脉的多样性,整合城市街区,同时设置塔楼,为城市社区提供地标。这个项目引人注目的玻璃雕刻形式使建筑的透明度最大化,同时反射周边的多样环境。自然光照亮了整幢建筑,给周边街道、景观和社区带来了活力。

地面层平面

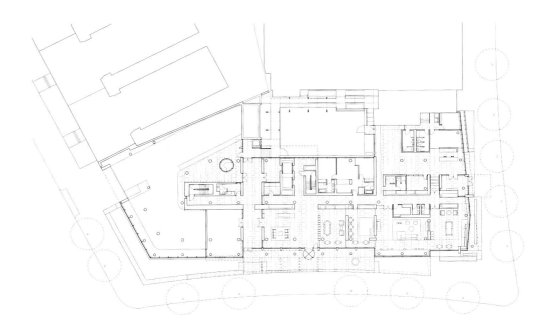

三层平面

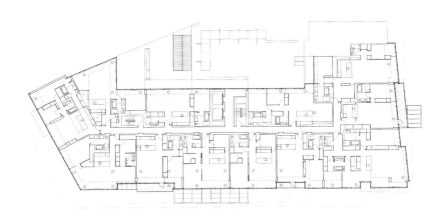

0 12 24 48

西立面

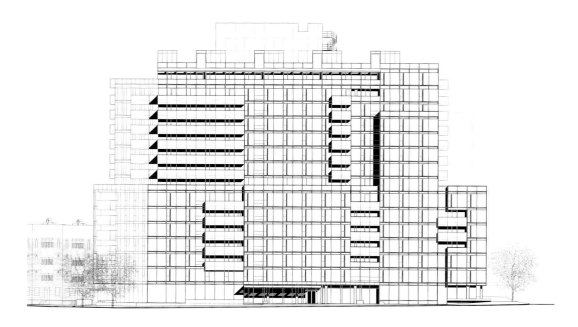

南立面

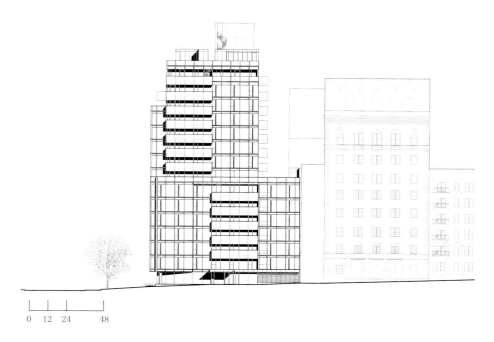

0 12 24 48

南佛罗里达住宅
Southern Florida House

南佛罗里达住宅
Southern Florida House

佛罗里达州棕榈滩，2004

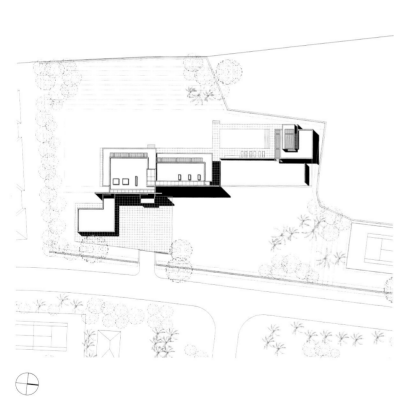

南佛罗里达住宅证实了迈耶持续地将精力投入到建筑的简约风格上。客户和建筑师共同的目标就是构建一个忠实于光影和空间原则的永恒建筑。通过密切合作，客户和建筑师共同完成了一个组织有序的设计作品，实现了客户的理想目标：不仅实现了夫妇俩对艺术的追求，同时也满足了有三个孩子的五口之家的实际需求。

住宅紧邻宁静的湖畔，沿着棕榈滩主要的林荫大道，景观设计运用了若干建筑学元素。入口庭院处实体的阶梯状垂直面实现了步移景易的效果，完美地将建筑与景观衔接起来。入口的坡道、主楼层的升起平台和悬浮的屋顶引导和构建了贯穿整个建筑的流动空间。楼板和墙体从建筑的结构中挑出，穿越了景观区域，以雕塑公园、游乐场、游泳池和池边小屋连接建筑。高耸的墙壁和与天花板等高的玻璃窗让建筑三面可观湖景。间断的屋顶、露台和开放空间使得光影在室内外产生戏剧性的变化。相互分离且具有层次的天窗走廊、入口门厅、核心服务区和主要的房间以动态而又简洁的布局方式组合。屋顶平面处于两个交错的玻璃体块之上，构成建筑的两部分，一部分包含了两层高的起居室、餐厅、厨房和二层的主人套房；另一部分包括一层的家庭活动室和图书室，以及二层的儿童房。

房子内外的细部小巧、精确，严格遵照当地法规的要求。建筑所使用的材料包括玻璃、钢、石灰石、金属板和石膏——这些都是白色调的（尽管它们各自有不同的纹理），彼此之间平衡、叠加。作为经典的理查德·迈耶住宅，尺度、比例、动静处理和空间关系都以低调而优雅高贵的方式处理。

几何

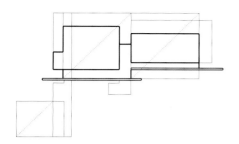

结构

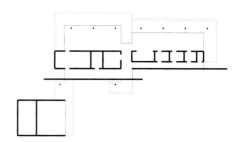

入口/流线

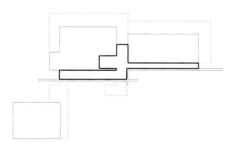

围护

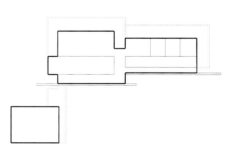

体积/层次

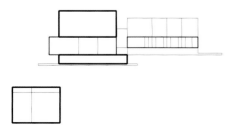

元素

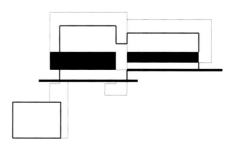

地面层平面

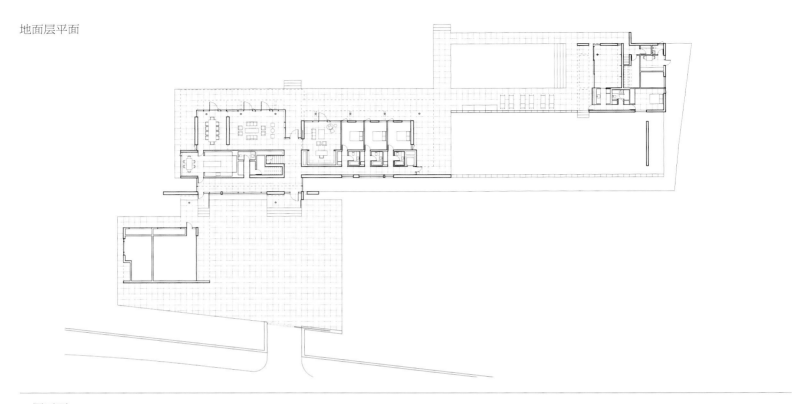

一层平面

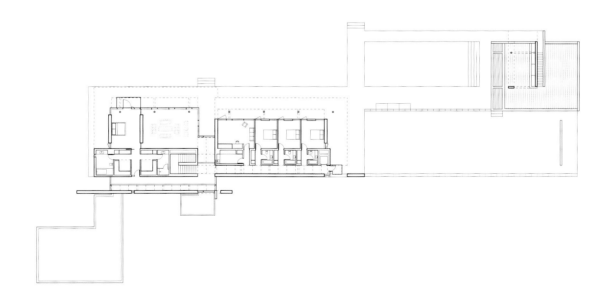

西立面

东立面

深圳住宅
Houses in Shenzhen

深圳住宅
Houses in Shenzhen

中国深圳，2005—2008

深圳项目的五个住宅为奢华的山间别墅，均被当作独一无二的作品来设计，在保持设计手法统一的同时，每栋住宅均采用了独特的处理方法。住宅的面积从600到1000平方米不等，它们的整体高度和楼层数根据场地情况确定。项目的目标是创造一个特殊的住宅邻里关系——与山体环境之间巧妙的平衡。建筑的选址和每个住宅的设计需要使私密性和观景面最大化，同时使对场地的影响最小化。

这块场地以陡峭的山体景观和山谷为特征，五栋住宅均坐落在陡峭的山坡上，沿着东西轴线，这样的布局方式使南向景观最大化，同时保证东西向有阳光。五栋住宅均从北面车库和车道的位置进入，穿越一个面积较大的正式门厅。每栋住宅的路线都是从不透明的北面走向通透的南面，露台和游泳池设置在可以提供最佳光照和观赏海景位置的地方。

两层通高起居室设置在可以获取东向和南向最佳视野的位置，是每座住宅的核心。餐厅和厨房毗邻起居室，是公共核心空间的重要组成部分。卧室设在比起居室高一层或是两层的楼层上，从公共空间分离出来以保证私密性，主卧拥有最好的视野，以及最佳的私密性。家庭活动室通常设在卧室旁，是温馨的生活空间。底层和地下室是娱乐室和服务空间。外部空间被建筑结构如南面的阳台和露台整合起来，悬挑和遮阳棚保护外部空间免受阳光和雨水的影响，从而全年均可使用。

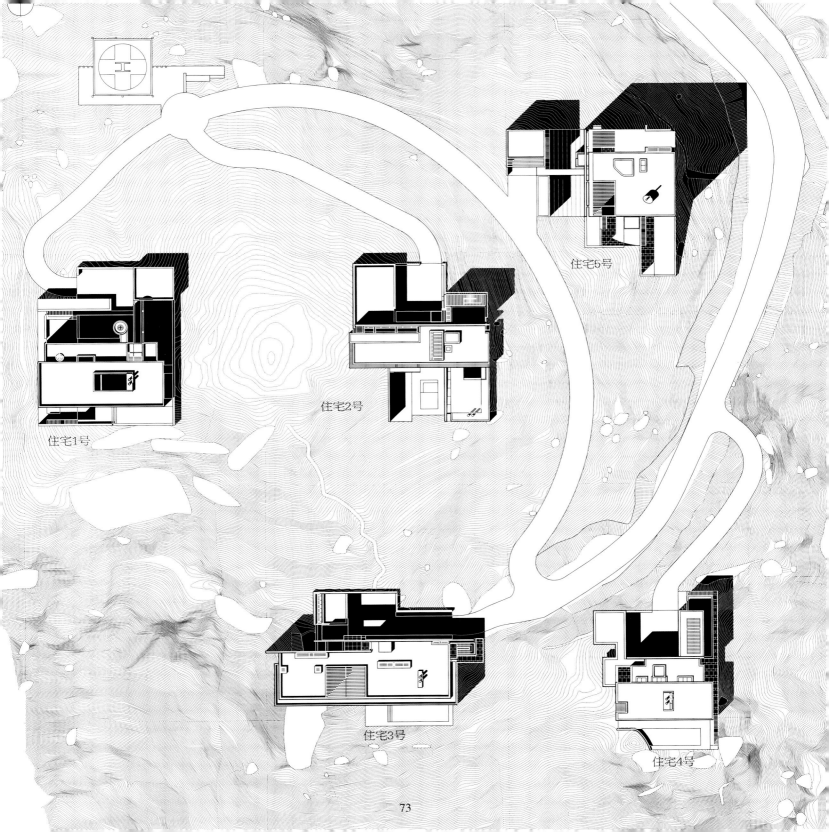

住宅1号
底层平面

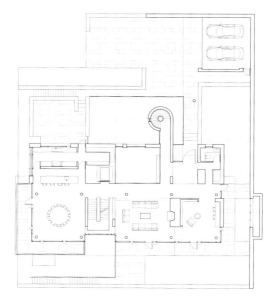

三层平面

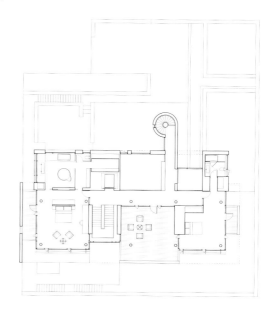

住宅2号

入口层平面

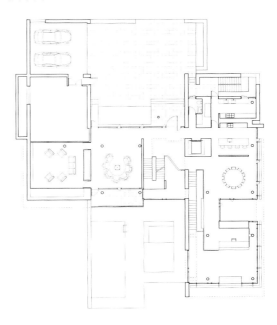

二层平面

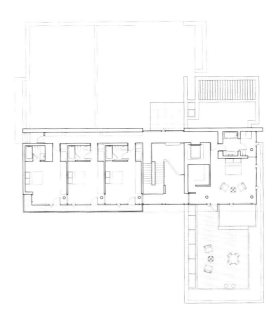

住宅3号
入口层平面

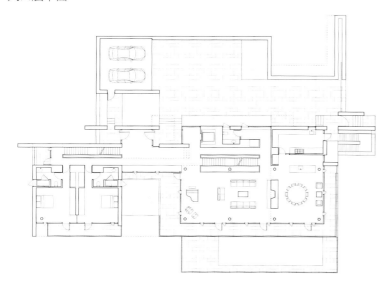

二层平面

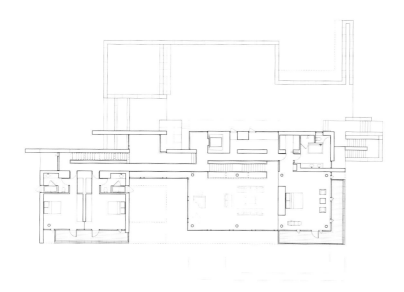

住宅4号

入口层平面

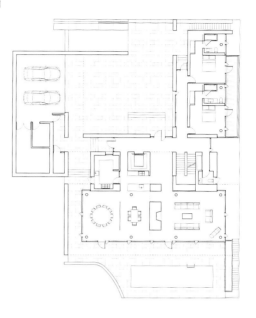

二层平面

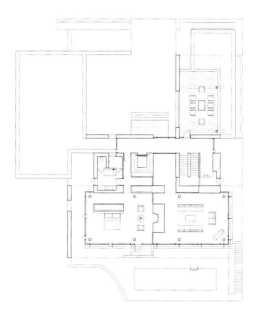

住宅5号

入口层平面

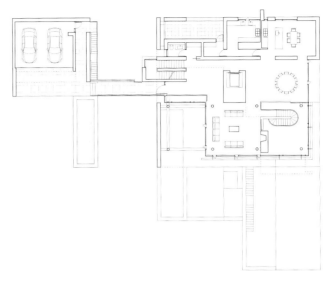

二层平面

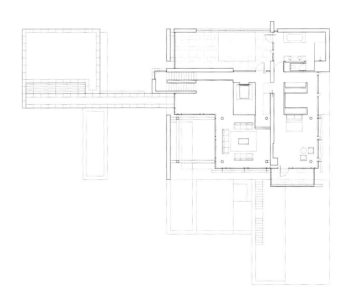

华侨城半岛别墅
地面层平面

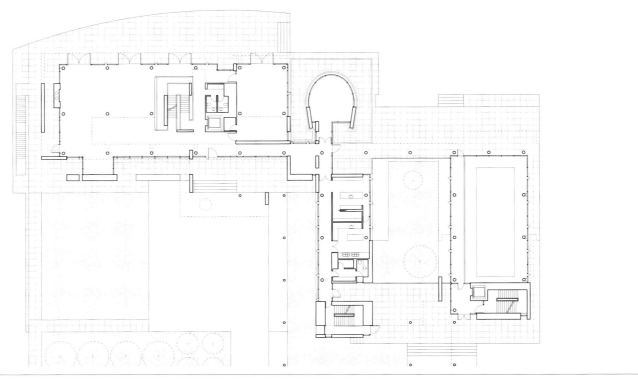

三层平面

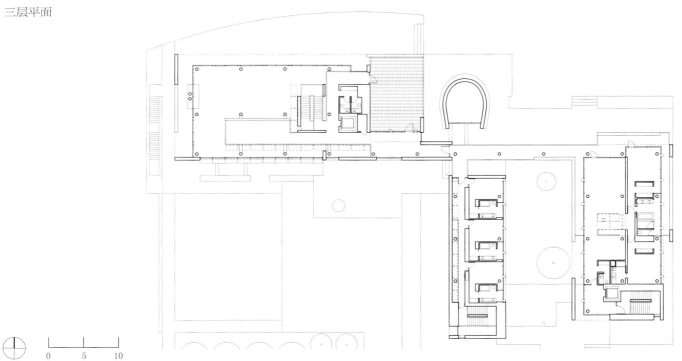

北立面

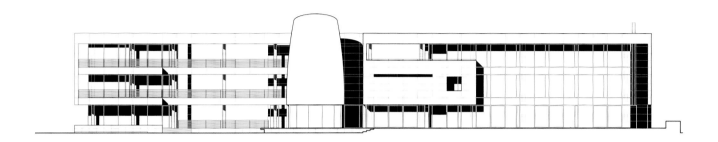

南立面

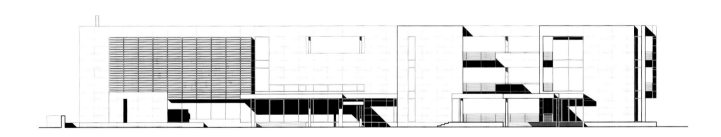

华侨城半岛别墅
OCT Peninsula Villa

瑞克麦斯住宅
Rickmers House

瑞克麦斯住宅
Rickmers House

德国汉堡，2002—2008

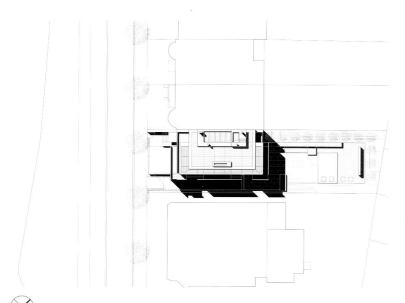

9米宽的联排住宅被分为四层，地面层为起居室，二层是客房和学习室，三层是儿童房，顶层为屋顶露台和阁楼，阁楼内为主卧和浴室。场地被挖方以获取空间设置工人房、储藏室和车库。

建筑拥有经典的联排住宅平面，被抬高的入口十分宽敞，可进入前院，与顶部采光的交通核位于一条轴线上。入口系统为通高空间，浴室被设置在等宽的服务核中，服务核内包含垂直交通。

建筑的三个自由立面被两层透明的白色玻璃遮挡，巧妙的开洞可引入景观和阳光。西面的承重墙在每一层都与独立的混凝土柱连接，这些柱子就在建筑的东立面里。后花园的铺地上设有玻璃底的水池，给下方的车库提供采光。

地面层平面

一层平面

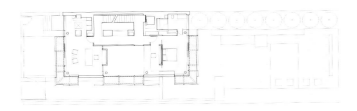

二层平面

三层平面

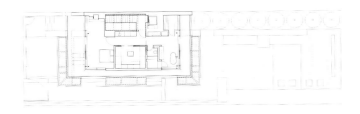

第五大道公寓
Fifth Avenue Apartment

第五大道公寓
Fifth Avenue Apartment

和罗斯·塔洛·梅尔罗斯住宅
纽约市，2007—2008

这个约232.3平方米的临时居所占据了第五大道标志性的雪莉酒店中的整整一层。设计改变了最初的布局，从而将处于中心公园东南角对面这一地理位置带来的自然光和视线的优势最大化。精心挑选的材料和简洁的现代细节设计重新诠释了标志性建筑的精致与优雅，创造了与周围环境共生的当代住宅。

公寓通过一组木材与玻璃组合的百叶门进入，这些门封闭了电梯前厅，建立了一个正式的入口。空间的层次感保证了室内空间的私密性，同时也让自然光渗入核心区域。进入入口大门之后，就是一个宽阔的开放空间，延长了沿着第五大道的塔楼的长度，通过定制的木制书架设计展示客户的非凡艺术品收藏。原有的三个独立房间被改造成一个宽阔的流动空间。这个空间里包含起居室，带有可伸缩电视的非正式座位区，还有一个餐厅，旁边是开放式厨房。延长的石膏墙可以展示艺术品，提供公园的全景视野，与空间的长边侧面相连。

这一公共区域连接了在电梯核对面的两个私密空间。北面，一条走廊连接办公室、服务入口和客房；南面，是带有两个完整盥洗室的主卧套间，这个位置采光最好。项目的黄金位置提供了得天独厚的光照和视野。建筑开放式的平面和抬升的天花板使这种光照和视野成为可能，同时也实现了塔楼含蓄的轻盈感。

地面层平面

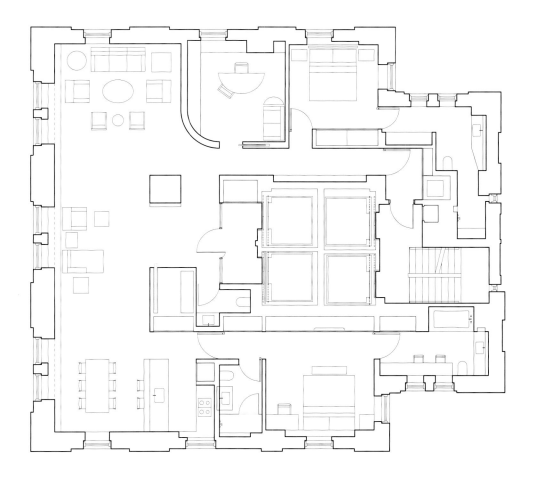

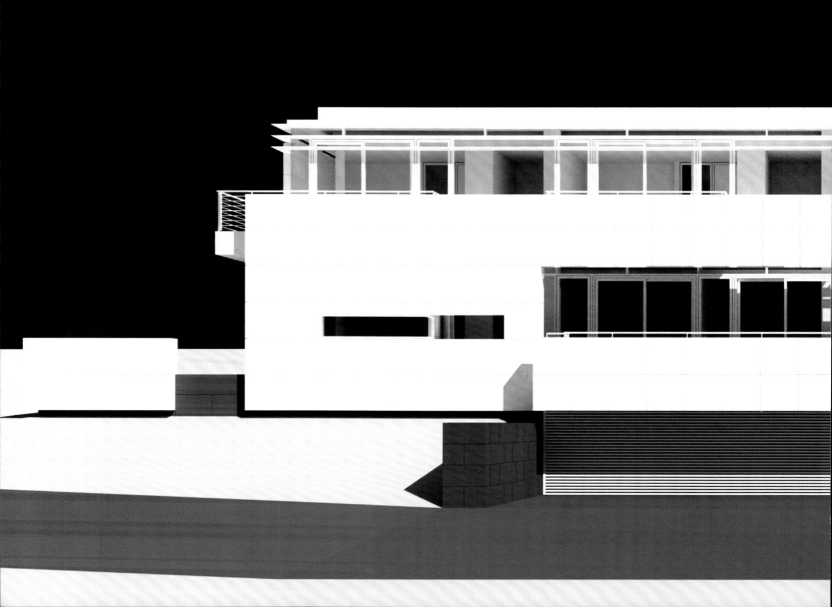

卢森堡住宅
Luxembourg House

卢森堡艾森博，2007—2009

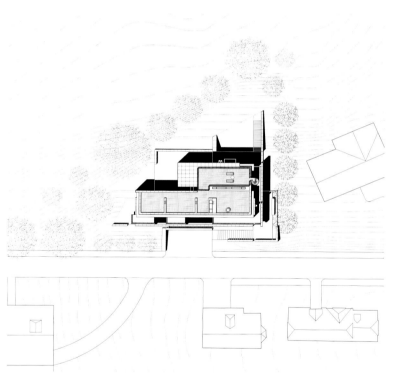

项目为私人住宅，场地位于农村，斜坡地形，树木繁盛，与世隔绝，周围景色可尽收眼底。建筑的主体就像一个玻璃棱镜，L形的布局将不同的空间分离。住宅的南面和东面被不透明的墙面包裹，保护建筑免受街道和邻近建筑的影响，同时将住宅锚固在景观中。主墙和玻璃体块之间的对话发生在空间间隙里，通过一连串露台为内外架起了桥梁，恰当的开口控制着光和景色的引入。整栋建筑被连接到楼梯间的交通核分为两个空间：一个沿着南面的实体区域，以混凝土结构基墩的韵律为特征；一个为北面的中空区域，由轻钢结构系统加固。玻璃墙将两个区域封闭，同时营造了动态空间的感觉，由于实体元素的存在既独立又均衡。

住宅的底层——包括功能空间和停车空间——被切割并嵌入场地的北侧，位于斜坡上的庭院一侧开敞，底层包括桑拿浴室、健身房和通向上层花园的入口。主要的公共空间均位于地面层，南侧是开放式厨房、客房和儿童游乐室，北面则是起居室和餐厅。次卧和主卧套间安排在南面的顶层，主墙在整个立面上隔绝了外部的干扰，也用玻璃百叶系统隔绝过多的阳光。这些房间的对面，是学习房、图书室和可以俯瞰起居室的休息室。

几何

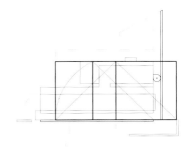

结构

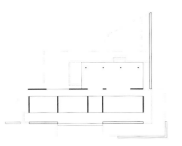

图解

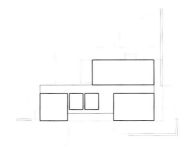

实/虚

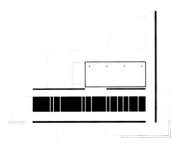

底层平面　　　　　　　　　　　　　　　　　　地面层平面

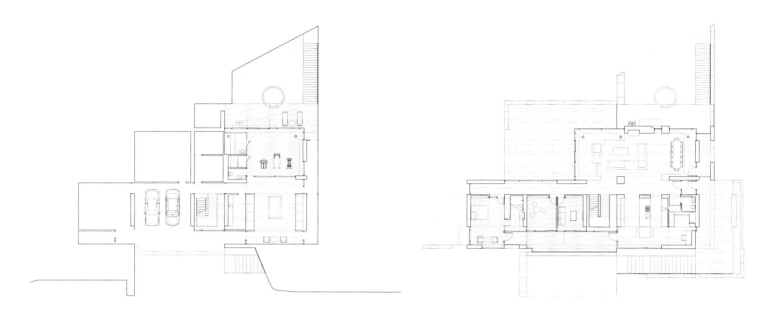

一层平面

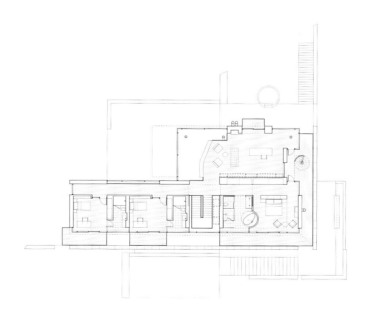

0　2　5

南立面

东立面

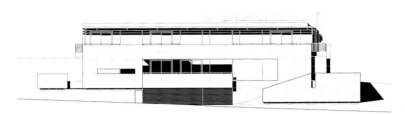
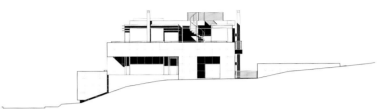

北立面

横剖面

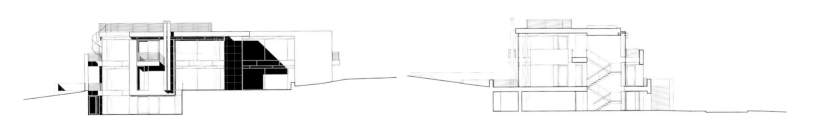

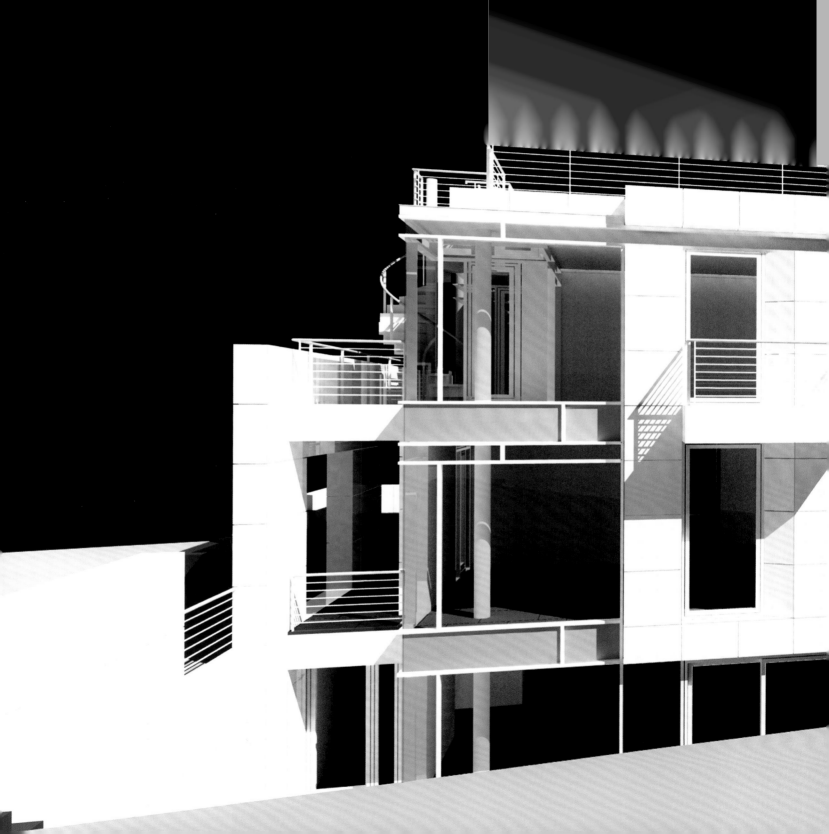

博得鲁姆住宅
Bodrum Houses

博得鲁姆住宅
Bodrum Houses

土耳其雅乐卡瓦克，2007—2010

项目坐落于土耳其博得鲁姆半岛雅乐卡瓦克小镇外的陡峭山坡上，由21栋住宅组成，可俯瞰雅乐卡瓦克海湾。整个场地地形充满戏剧性，其每一地块（约合0.4公顷）都十分独特，同时为住户提供隐私保护，遮挡来自毗邻地块的视线。建筑师提供了五种类型的住宅，每种类型住宅约330平方米外加一个40平方米的客人房。住宅包括独立的车库、游泳池和有凉台的小屋，规整地设置于裙房之上，以便让住宅的所有体块保持紧凑。

除了在场地中的位置不同外，每种类型的平面格局不变，裙房的组织方式依据场地特征改变。所有住宅的位置都能获取最宽广的视野，并且不论处于哪块场地，都能建立入口序列来进一步开阔视野。

每种原型均具有明确的步行序列，以引导观者进入外部入口楼梯，然后进入室内门厅，最后进入通高的起居室。每块场地都要求它们的住宅有二层入口，但空间体验序列是相同的。建筑内壁炉是核心的组织要素。每栋住宅都包括：地面层的起居室、餐厅、厨房和化妆室；一层的3个卧室；地下室的多媒体室、洗衣房和3个工人卧室。

建筑着力运用简单的处理方法：材料为石膏饰面的现场浇筑混凝土和大面积的玻璃。建筑内部运用了更加精致的材质，包括石材和硬质木地板。丰富的天井照明有效地形成了建筑的第五立面。建筑师给了建筑立方体的外观，住宅成为景观中的独立个体。因此，要产生大面积的退台，外部空间从建筑体块中切割出来，同时掩于屋檐下，因此每栋住宅均运用减法使建筑呈现出雕塑的品质。

1号住宅
一层平面

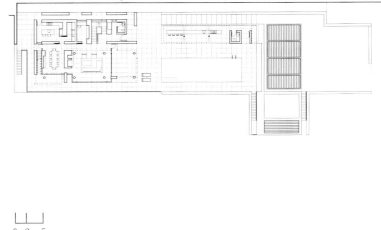

0 2 5

轴测图

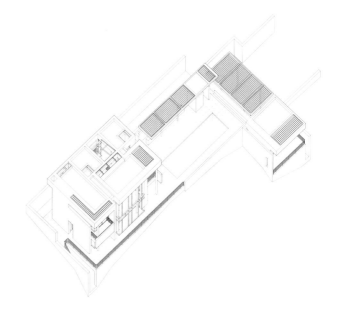

2号住宅
House 2

2号住宅
一层平面

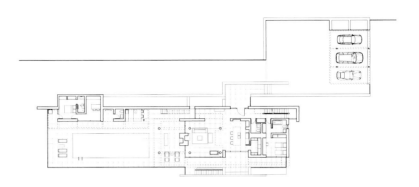

0 2 5

轴测图

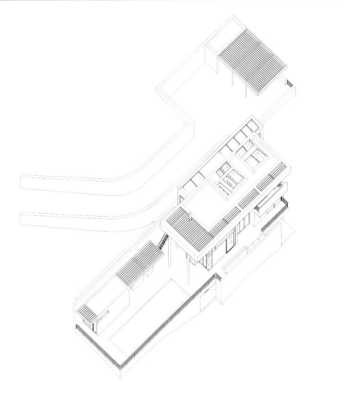

3号住宅
二层平面

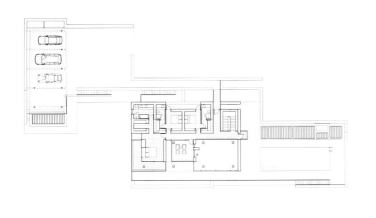

轴测图

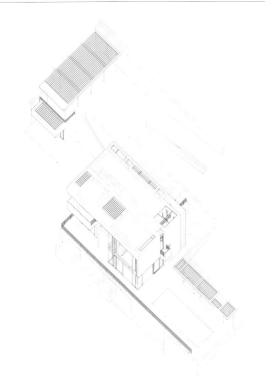

4号住宅
一层平面

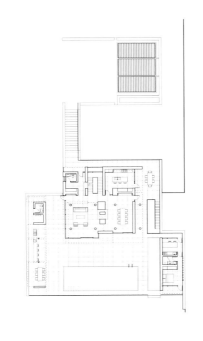

0 2 5

轴测图

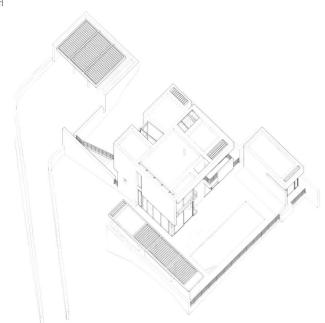

5号住宅
二层平面

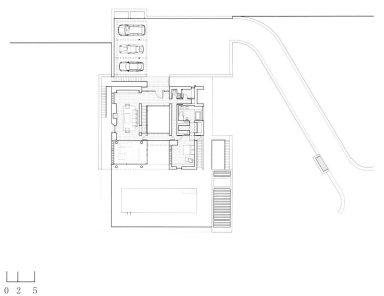

0 2 5

轴测图

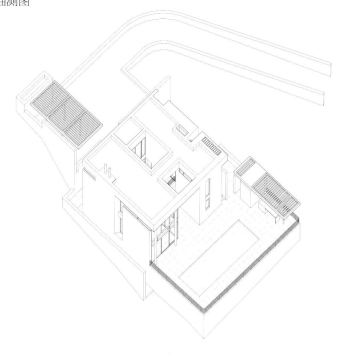

天津别墅
Tianjin Villa

天津别墅
Tianjin Villa

中国天津，2009－2012

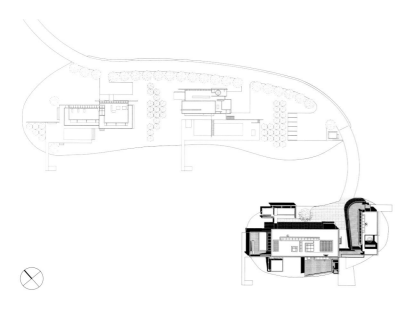

本案为私人住宅，坐落在称为"世界"的天津星耀五洲住宅区内，与酒店分离。项目要同时提供公共功能和私人住宅功能，以及为客人提供服务设施，包括室内游泳池，总面积约为3500平方米。白色的面板和玻璃的表面并置，形成了起伏的形态，具有很强的几何体量感。一、二层空间容纳娱乐和商业功能，适应于各种规模的聚会。门厅引导拜访者进入正式的起居空间，光从两层高的玻璃进入起居空间。门厅中设有雕塑感很强的楼梯，连接着别墅的三层空间。三层为别墅的私人区域，包括家庭用餐区、厨房、儿童房、家庭活动室和一个大露台，露台上有一片竹林。主卧套房位于顶层，可获得全景视野，并可从一个内部的开洞俯瞰两层高的家庭活动室。

场地的南面，宁静简约的茶室漂浮在无边的水面上，与倒影相映成趣。室内游泳池和客房为主要房间，并且把场地固定在岛屿的东端，同时和前院的入口相连。整体覆盖的百叶窗过滤了射入室内游泳池的光，光随着一天的时间变化不断改变着空间效果。客房，连同SPA设施一起，占据了游泳池的上面一层，业主可从外部楼梯进入或者从建筑主体二层的封闭连廊进入泳池。

一层平面

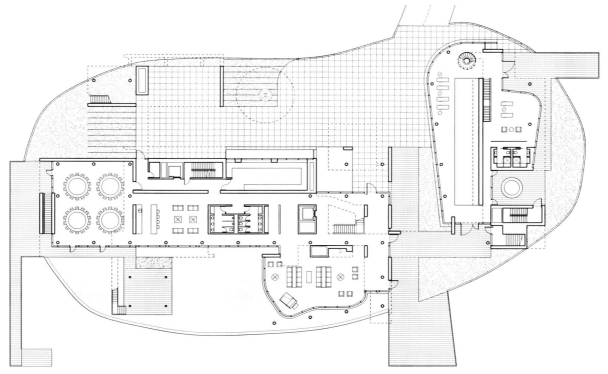

二层平面

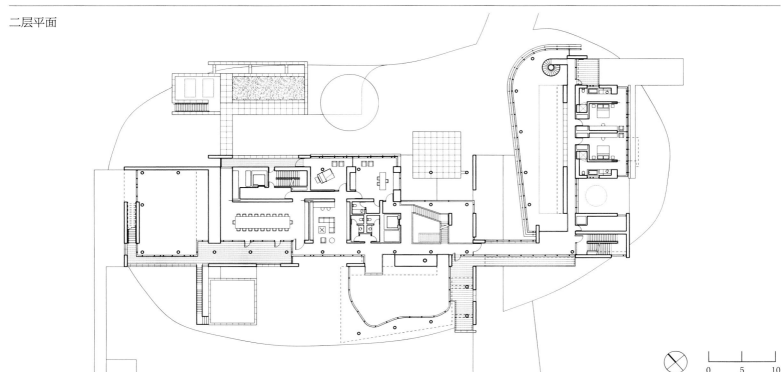

南立面

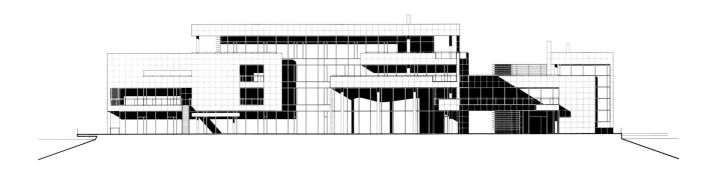

北立面

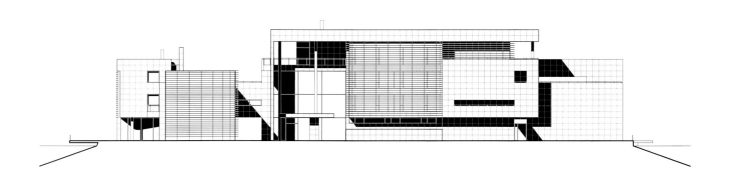

横剖面

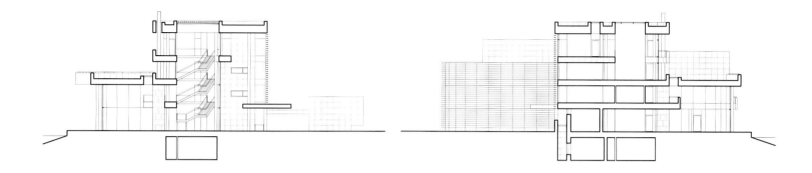

纵剖面

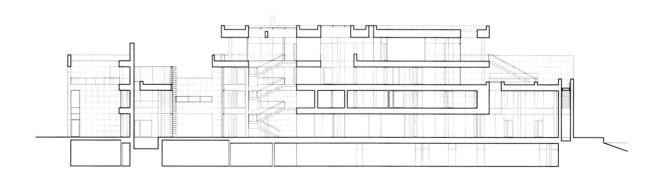

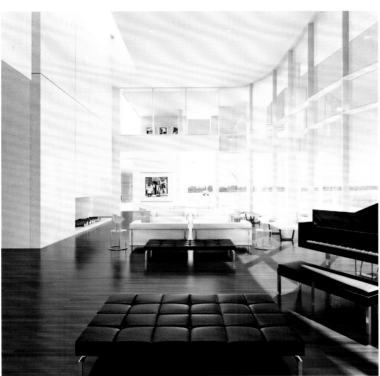

公共建筑与项目
Public Buildings and Projects

阿尔普博物馆
Arp Museum

阿尔普博物馆
Arp Museum

德国雷马根-罗兰泽克，1978—2007

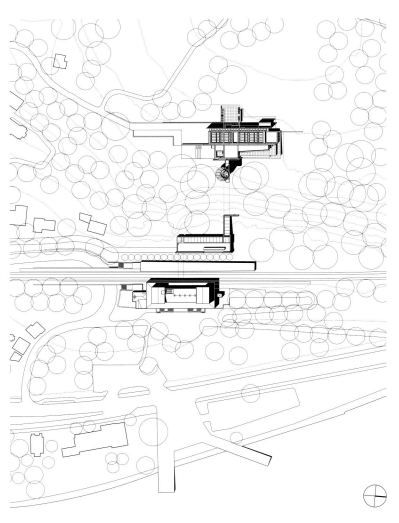

阿尔普博物馆的设计展示了两个因素的完美结合：一是莱茵河畔的壮美场地，二是博物馆的使命——展示达达主义大师汉斯·阿尔普的作品和他的生平。建筑的入口序列不是从博物馆本身开始的，而是始于山脚下河岸边一个古老乡村的火车站，火车站从20世纪60年代开始就被用作展览空间，车站延伸至40米长的地下隧道，衔接铁轨和独立于主体展馆的展览廊道。除了提供配套的临时展览空间，展馆还建立了一种期待感和不确定感，这种感觉在下一个体验序列里被进一步强化。接下来，体验序列物化成地下隧道，终点是富有戏剧性的40米高的传动轴，在那里可以进入两个玻璃电梯。电梯通过传动轴上升到顶层的锥形塔结构。在塔楼的顶端，电梯通往一段长达16米、被玻璃封闭的桥，代表进入博物馆体验序列的高潮阶段。

博物馆一层入口的右侧是连接上下层的独立楼梯，左边设置了可以俯瞰低一层大厅的洞口。大厅为参观者提供休息空间，底层是教室、行政办公室、服务设施，以及运输、接收艺术品区域的入口。上层是两个大型展廊，占据了一个由柱子支撑、看似自由漂浮的平台，让参观者忽略了地面层东西两边的展廊。上层的主要展廊几乎被玻璃天花覆盖，十分光亮，与一系列0.6米宽的可调节铝百叶一起，提供全自然或者人工调节的光环境。覆盖了两层高的面向莱茵河的玻璃幕墙，为博物馆开启了整个观景面，幕墙类似百叶系统，让建筑获取令人震撼的山谷景观。

几何 入口

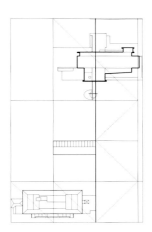 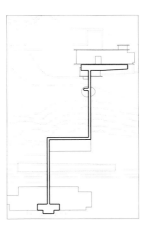

结构 流线

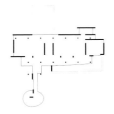 

元素 图底关系

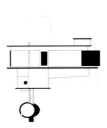 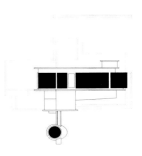

展览馆层平面

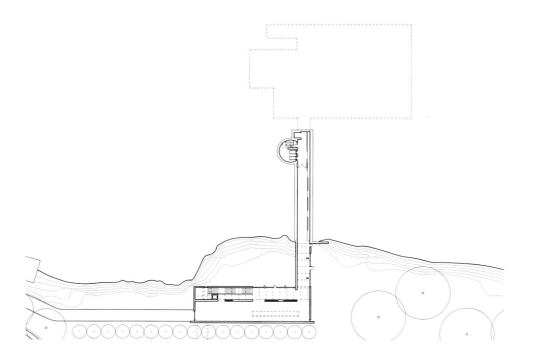

地面层平面

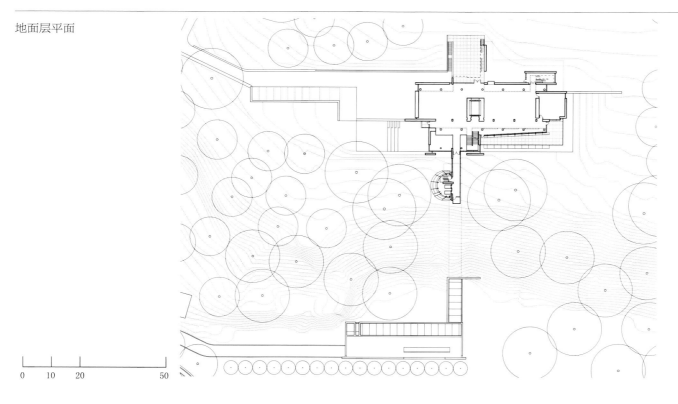

0 10 20   50

底层平面

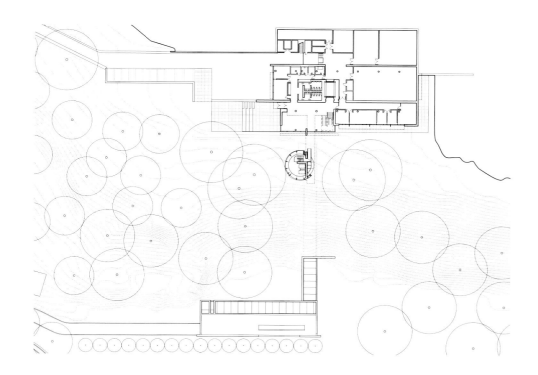

上层平面

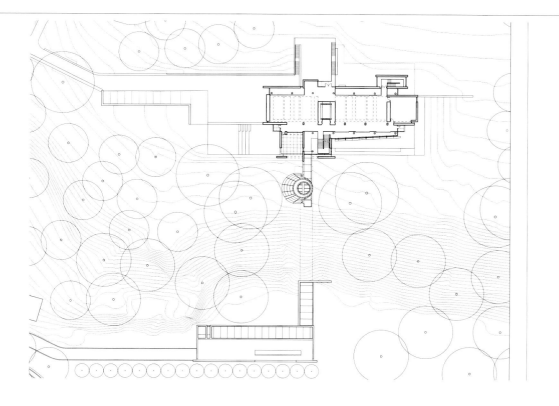

东立面

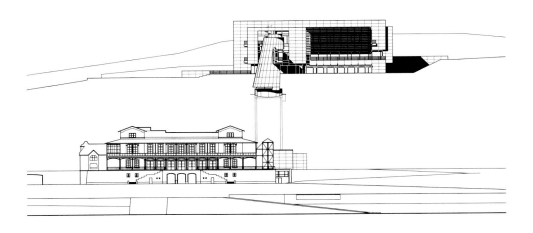

南立面

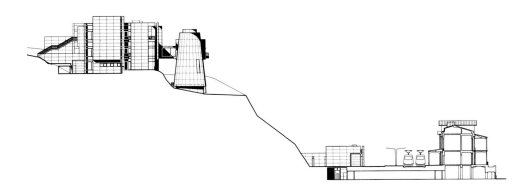

纵剖面

横剖面

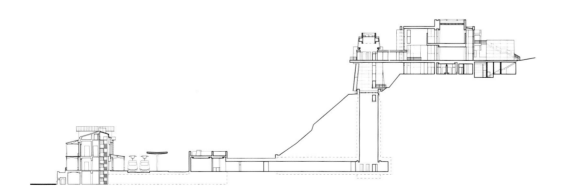

和平祭坛博物馆
Ara Pacis Museum

和平祭坛博物馆
Ara Pacis Museum

意大利罗马,1996—2006

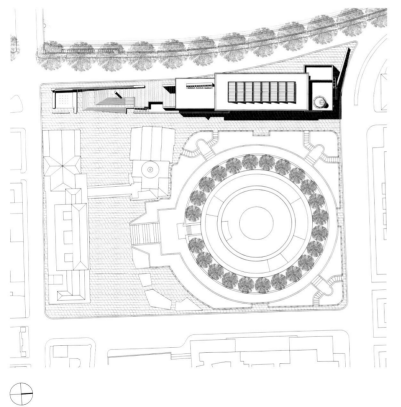

本案是台伯河岸边的和平祭坛博物馆的更新设计,该祭坛是一个可追溯到公元前9世纪的祭坛,现在位于奥古斯都大帝广场的西部边缘。本设计是保护罗马文化遗产工作的一部分,新的结构取代了原本就在衰退状态的古迹外壳。单层玻璃长廊建于较低的台基上,这个新结构在台伯河堤和建于公元前28年的奥古斯都圆形陵墓之间提供了一个透明的屏障。

祭坛在1938年墨索里尼时期从战神广场迁移至此。本案运用了一个调节线系统,让祭坛现在的位置与原始位置关联起来。将目前陵墓中心和原先场地之间的距离等分,产生了一个四方的城市网格,这个匀称的结构重组了广场及其周边环境。方尖碑作为历史的参考设于祭坛的南北轴线上。

清晰的体量和建筑的比例均参照古罗马建筑的形制。新建筑的最显著特征是近46米长、12米高的玻璃幕墙。七根细长的柱子构成了非对称的入口大厅,柱子由钢筋混凝土外加打蜡的大理石石膏筑成。大厅引导人们进入主厅,那里放置着和平祭坛。入口空间充分运用顶部照明形成柔和的光线,与严格对称的主厅之间形成对比,促成了一种自然的连续循环。主厅的屋顶被四个柱子支撑,顶部开有天窗,自然光被最大限度地利用,力图消除"虚假光影"。主体结构之外,低矮的石灰墙从主厅沿着古老的台伯河岸延伸出来,而观众席上的室外屋顶平台是博物馆流线的重要组成部分,它包括连续的酒吧和咖啡馆,在那里,可以观赏东面奥古斯都大帝广场和西面台伯河的风光。

底层平面

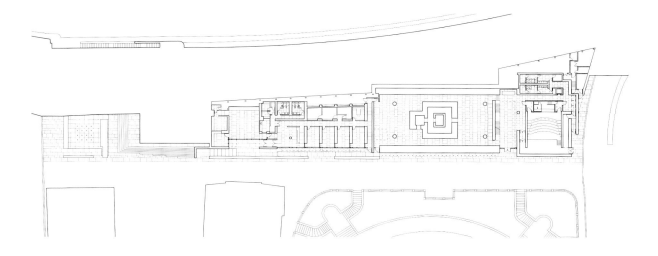

地面层平面

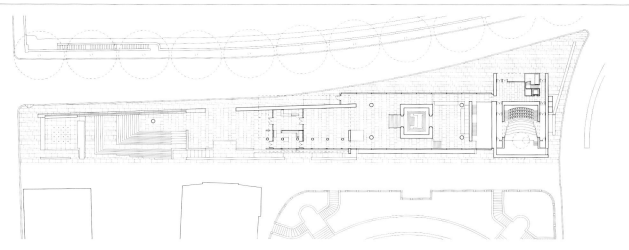

天窗层平面

东立面

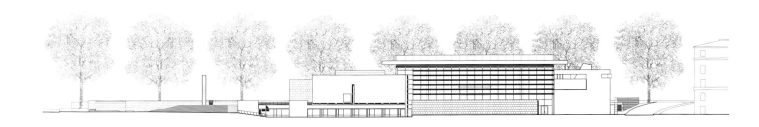

西立面

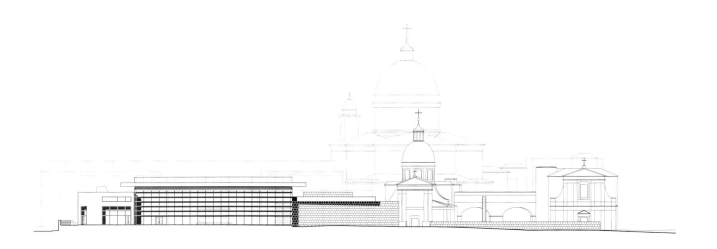

纵剖面

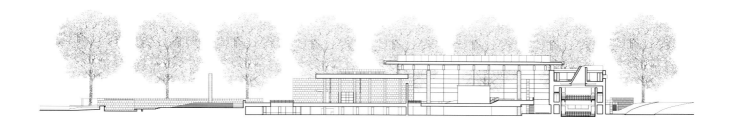

横剖面

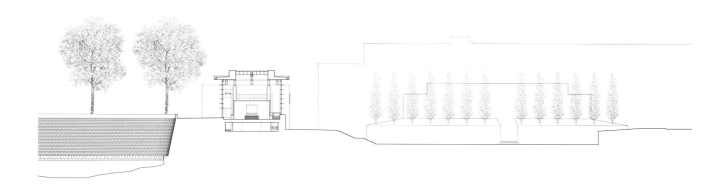

圣何塞市政厅
San Jose City Hall

圣何塞市政厅
San Jose City Hall

美国加利福尼亚圣何塞市，1998—2005

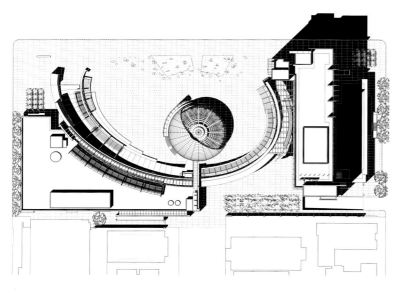

圣何塞市政厅在七块重建区的中心地块内，重建区内的新、老建筑通过街道、步行街广场、庭院和喷泉统一起来。周边建筑将会包括新的演出大厅、小学校、重建的教堂、城市-大学联合图书馆，还有两个公共停车场。该项目包括18层高用以容纳城市职能部门、城市理事会会议厅的办公综合体、大型文娱圆形大厅、外部广场和地下停车场。

这一意义深远的公共广场的核心是透明的穹顶入口，它既是市政府的标志，也是建筑主入口的标志。玻璃圆形大厅，虽然运用了更加现代的形式和材料，但体现了对传统穹顶的记忆，还可容纳大型公共活动，例如演讲、音乐会、展览。弯曲的墙壁定义了主厅，也限定了圆形大厅的空间，它的存在将建筑设备也整合其中。

圆形大厅是东侧高层办公楼的有益补充，高层办公楼里是城市职能部门以及与广场同一标高的许可中心，大厅西边是三层的委员会办公处，设有市议会室、公共会议室、零售店和其他部门办公室。圆形大厅内有一个巨大的楼梯，引导人们进入位置显著的市议会室，也可进入沿广场墙壁布置的外部楼梯。

环境

结构

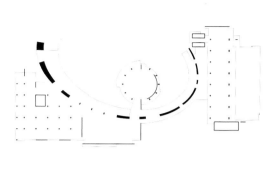

入口

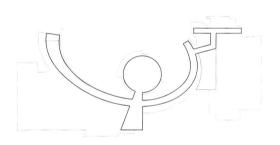

流线

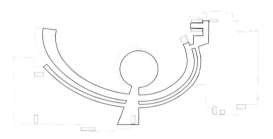

围护

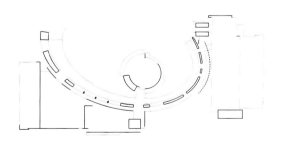

图解

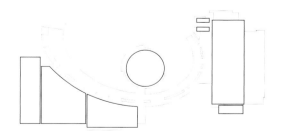

广场层平面

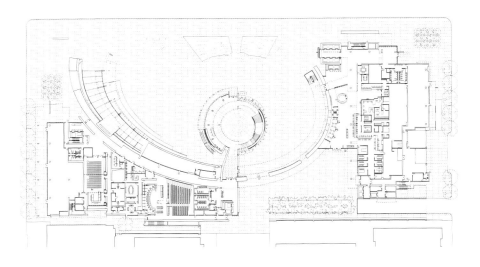

二层平面

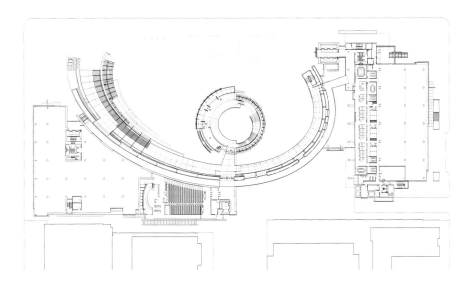

北立面　　　　　　　　　　　　　　　　　　　　　　　　　　　　　　　　　　　　东立面

南立面　　　　　　　　　　　　　　　　　　　　　　　　　　　　　　　　　　　　西立面

加州大学洛杉矶分校
艾利与艾迪斯·布劳德艺术中心
Eli & Edythe Broad Art Center, UCLA

加州大学洛杉矶分校艾利与艾迪斯·布劳德艺术中心
Eli & Edythe Broad Art Center, UCLA

加利福尼亚洛杉矶，1999—2006

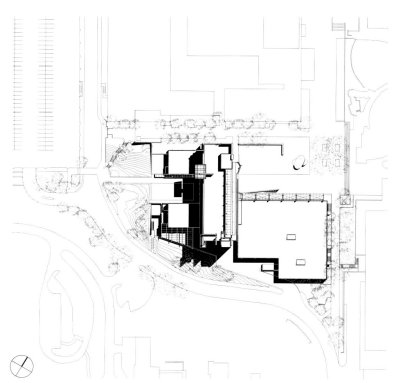

现存的约1.3万平方米的迪克森艺术中心位于加州大学洛杉矶分校校园内，始建于1964年。该项目为艺术与设计学院设计，提供数字艺术中心的扩建空间和普通教室。1994年北岭市地震期间，建筑损坏，结构需要加固。布劳德艺术中心在南北轴线的北端，是校园新北门的一部分。

教学工作室是艺术和设计课程的核心，这里是学生创作、与其他人分享设计的空间。利用类似阁楼空间的平面，融入灵活的空间和功能，设计区域的重新配置能更好地满足艺术学校的要求。为了发挥这种潜力，建筑师对艺术学校塔楼进行了彻底的翻新设计。外置结构拱加在了塔楼的西端，对现有结构框架进行了提升。翻新的空间设置了室内剪力墙的替代结构，让室内空间不受结构划分的限制，变得更加灵活。西侧的新的扶壁成为新服务电梯的围护结构，电梯将卸货平台与各个楼层和屋顶区连接起来。

原有庭院被利用来创造新的艺术表演空间，空间内进行了可适应不同用途的声学和光学设计。新的可穿越的阳台是流线的组成部分，提供了去往设计区域的通道，也对南立面的阳光进行了调控。水平叶片的格栅和屋顶为这些外部走廊过滤阳光并提供观景的可能。新的非正式会议空间可供教师和学生进行小范围的讨论，每一层的电梯核旁都设置了这样的空间。

结构

围合

流线

图解

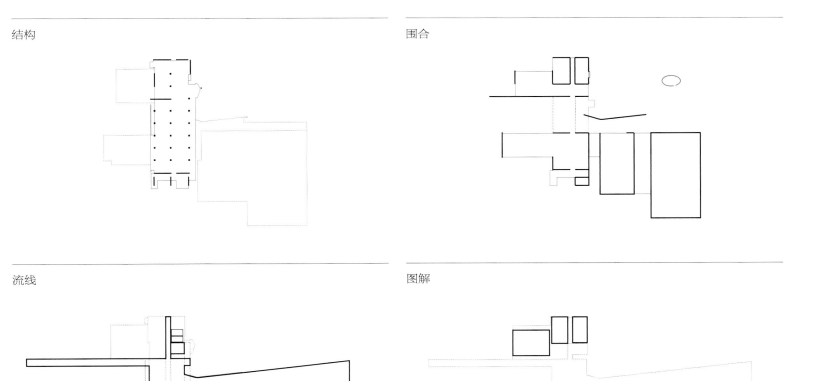

地面层平面

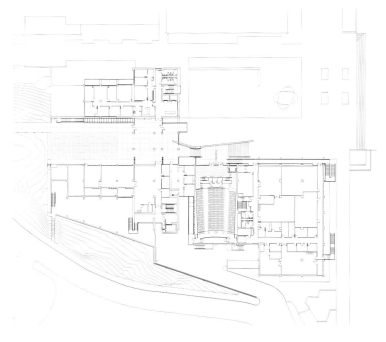

一层平面

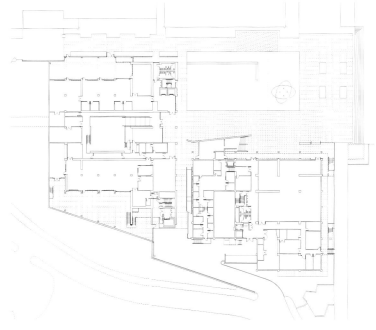

二层平面

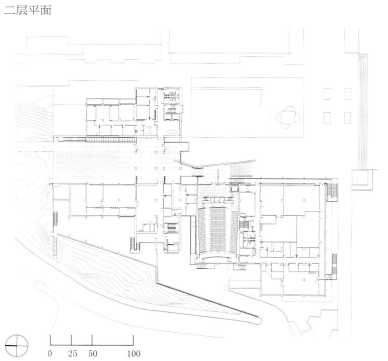

标准层平面

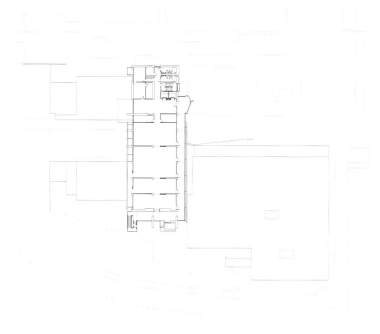

北立面

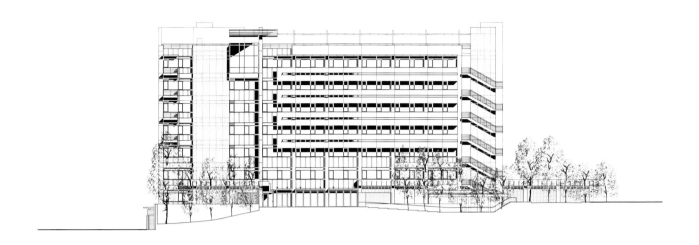

南立面

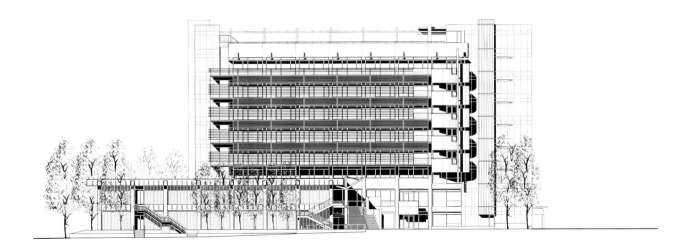

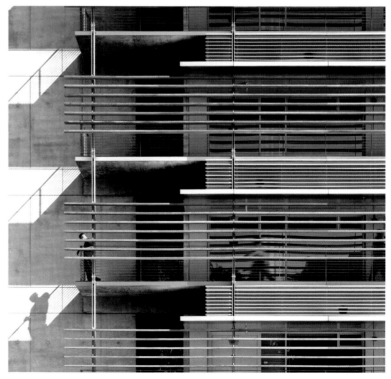

布尔达收藏博物馆
Burda Collection Museum

布尔达收藏博物馆
Burda Collection Museum

德国巴登巴登市，2001—2004

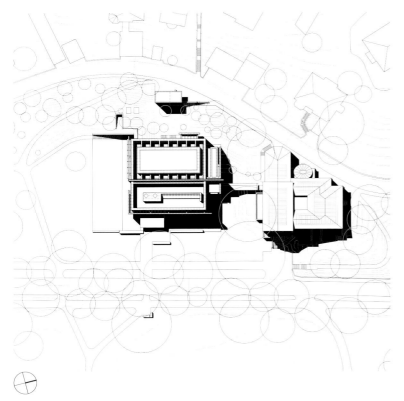

这座博物馆里是弗里德尔·布尔达收藏的20世纪古典、现代和当代艺术品。设计试图融入里奇腾塔勒大道的景观环境，同时与邻近的博物馆经典的建筑轮廓相协调。整个建筑的形式和比例与邻近博物馆的抬升底座和柱顶位置相似，但是每个构件均保持自身的建构特征。

坐落在公园繁茂的树林中，三层的博物馆从东侧的门廊进入。在二层，玻璃桥连接本案与邻近的博物馆。连接桥的细部设计精巧细致，以便尽量少地对现存博物馆的特点产生影响。

进入博物馆后，观者右转穿过大厅/接待区到达横向的四层斜坡大厅，斜坡大厅和一层之上连接现有博物馆的桥位于同一条轴线上。加上相邻的电梯，垂直交通作为主要交通方式可到达次要展廊（悬挂在地面层展廊之上）的入口，也提供了通达位于低层和俯瞰入口夹层的辅助展览体块的入口。

上层的主要展廊跨越了整个建筑，因此给人一种漂浮的感觉。它由斜坡大厅进入，经过一个可观赏周围公园或者建筑下部空间的桥。上层展廊的嵌入式楼板和下层展廊的边界墙让自然光渗透进入建筑下部的楼层。

几何

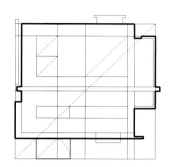

结构

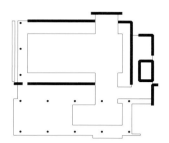

入口

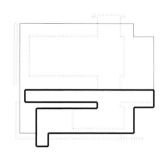

流线

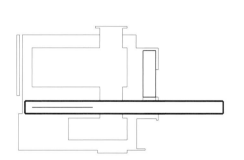

围合

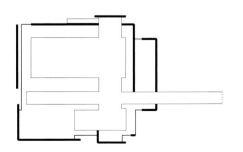

层次

地下室平面

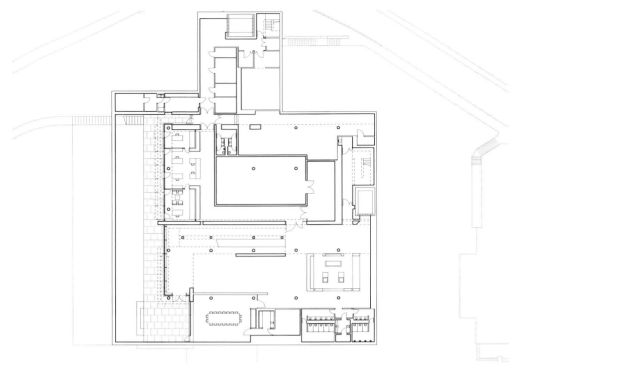

夹层平面

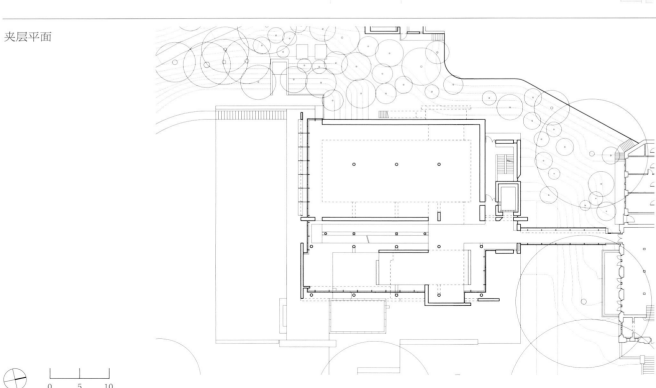

地面层平面

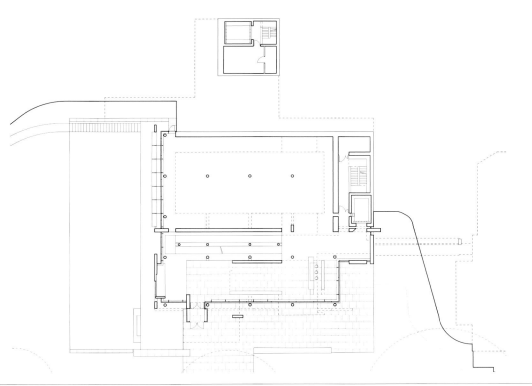

一层平面

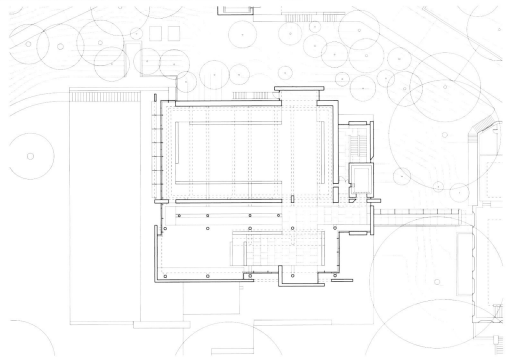

东立面

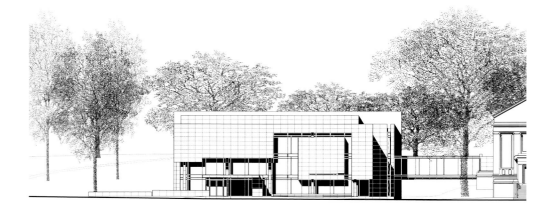

南立面

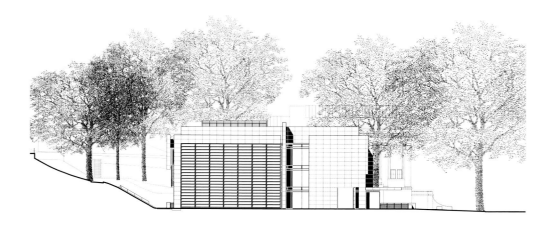

往西看的剖面

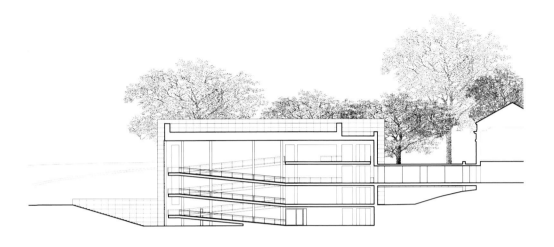

往北看的剖面

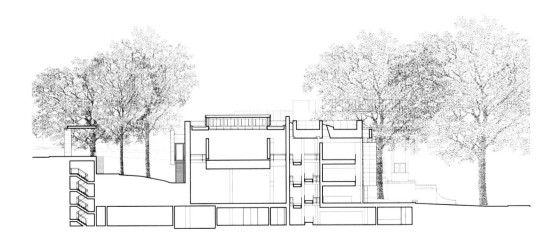

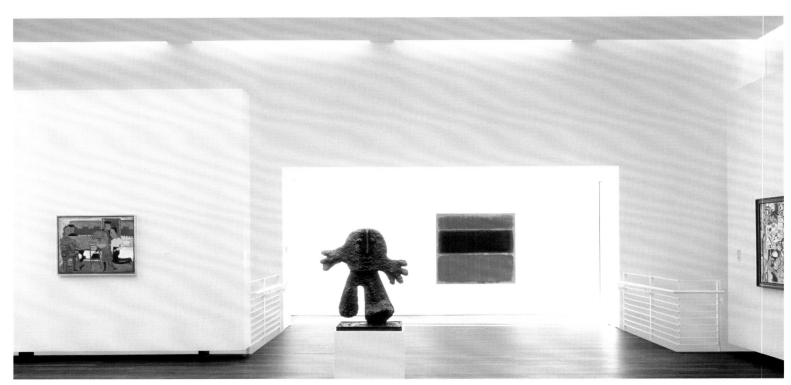

皮克-克洛彭堡百货公司
Peek & Cloppenburg Department Store

皮克-克洛彭堡百货公司
Peek & Cloppenburg Department Store

德国曼海姆,2001—2007

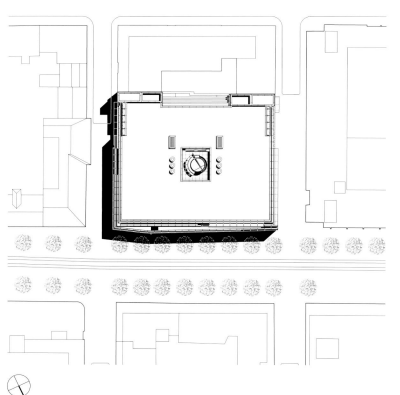

皮克-克洛彭堡公司的连锁零售百货公司设计旨在呼应曼海姆悠久的城市网格,此外提供城市景观中体现当前发展的地标。建筑仁立在城市主要干道边的突出位置,在博朗肯大道上的Stephanienstrasse与Kufürstenstrasse之间。两个街区的西北方阅兵广场具有很强的仪式感,坐落在Kurpfalzstrasse和博朗肯大道交口(实际上,这是巴洛克城市主要的南北和东西轴线)。建筑的弧形立面沿着步行和购物的主要轴线巧妙地悬挑出来,强化了皮克-克洛彭堡百货公司的形象。

并列的两个主要元素构成了建筑外表的标志:一个是沿着零售区轻盈透明的玻璃幕墙;一个是用罗马石灰石作为外饰面,容纳垂直交通和机械送风系统的庞大核心筒。三层高的玻璃围护结构从建筑主体悬挑出来,好像漂浮在博朗肯大道之上。透明的玻璃带形成了沿楼板设置的百叶窗,创造了一个通风的屋顶结构。

建筑通过一系列微妙的体块调整与周边城市环境形成互动。考虑到周边建筑的高度,顶层的房间稍微缩进。此外,大楼通透的特点强化了公共空间的属性。空间的流动将公共领域延伸到半公共领域。连续空间以一个悬臂式的屏幕作为结束,屏幕反射了人行道和市场内人的活动。建筑里容纳了五层零售空间,顶层为管理空间。圆柱形的天窗有日晷的元素,日晷可实现光影的连续变化。

几何

结构

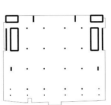

流线

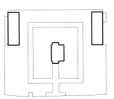

围合

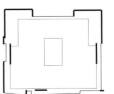

标准层平面

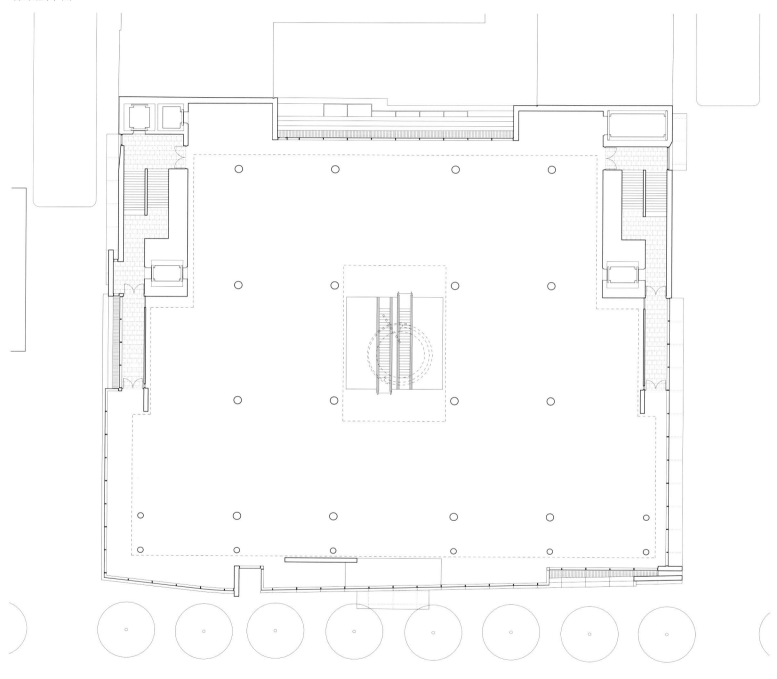

北立面

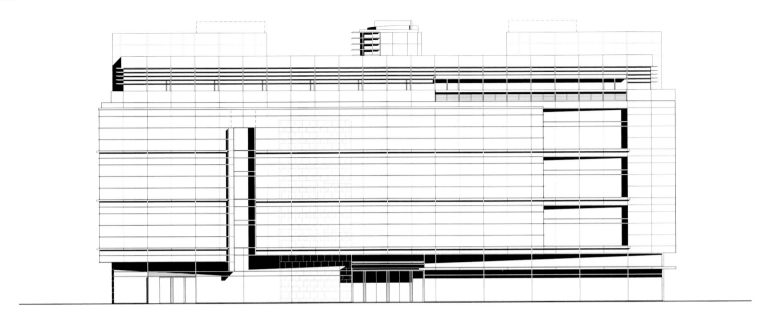

西立面

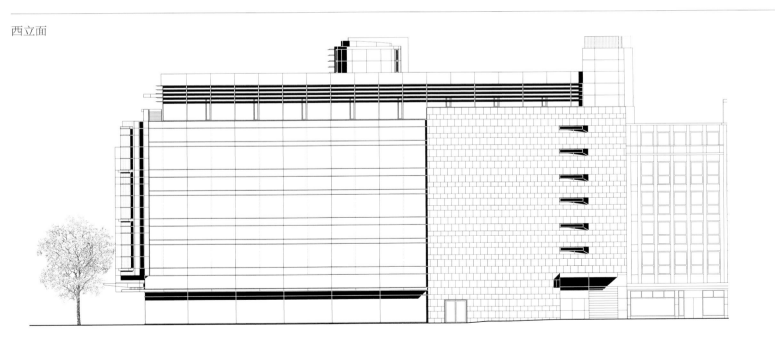

康奈尔大学威尔大厅
Weill Hall, Cornell University

康奈尔大学威尔大厅
Weill Hall, Cornell University

纽约伊萨卡岛，2001—2008

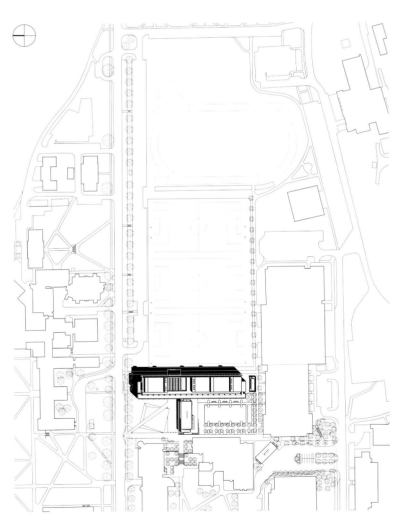

威尔大厅是康奈尔大学基因组的基石，跨学院、教师组织的研究、开发、教育计划旨在保持康奈尔大学在生命科学研究领域的主导地位。本案融合了生物、物理、工程和计算机科学各领域，努力加深在社会、法律、伦理和商业方面对基因组学的理解。

威尔大厅支持来自全国各地的大学教师、学生和科学家之间的研究合作。建筑总面积2.44万平方米，容纳了研究和教学实验室、基因组技术服务室、远程教学中心，并在楼上4个楼层设置"企业孵化器"，此外还有大型地下室，内有植物园和控制植物环境的相关设备。

威尔大厅在多功能运动馆的西边，形成了一个鲜明的入口，同时整合了环境中的现有科学设施，形成了场地整体关系。建筑组件通过分层结构组织，体现了一种线性的组织机制，通过利用风景和自然光提升内部空间的品质。面向多功能运动馆的立面重新定义了开放空间，带来了美国康奈尔大学中央校区的新形象。设计的创新方法强调设计和施工过程中的可持续发展和保护，使得建筑赢得了LEED金牌认证。节水景观和屋顶绿化设计显著减少了热岛效应的影响。生态屋面覆盖了建筑屋顶的50%，它能吸收雨水并提供隔热保护。高效率机械系统的设计，预计节约的能源能超过行业标准40%以上。建筑中运用了调节光、温度和风速的探测器，以及可减少光污染、提高水资源利用率和降低物质排放的系统。在施工过程中，物质循环、室内空气质量管理、地域材料的使用和经过认证的木材被优先考虑。总体而言，威尔大厅比同类建筑每天减少了30%的能耗。

| 几何 | 入口 | 围合 | 垂直/平行 |

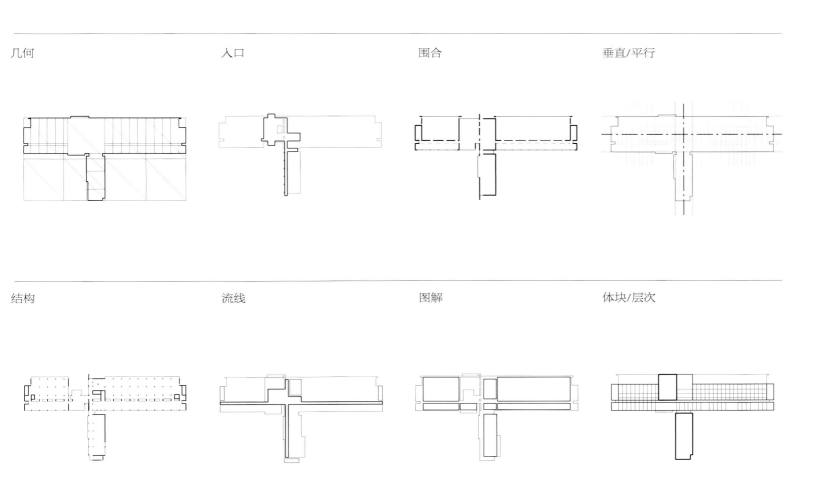

| 结构 | 流线 | 图解 | 体块/层次 |

轴测图

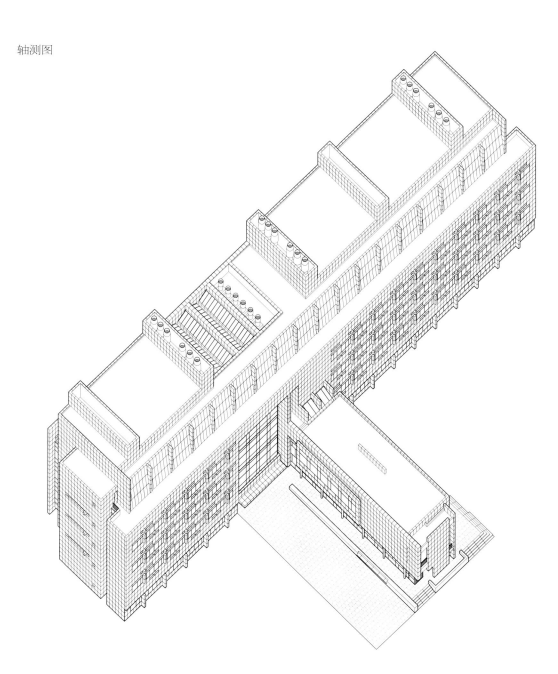

地面层平面

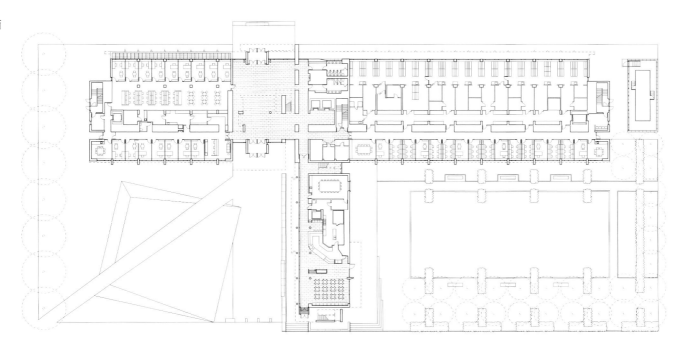

二层平面

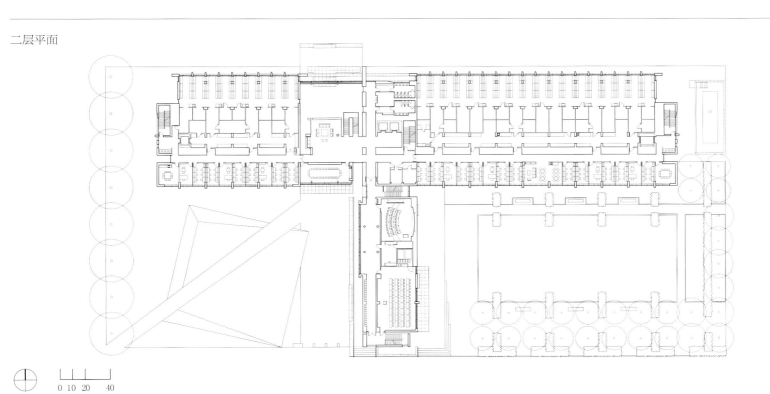

东立面

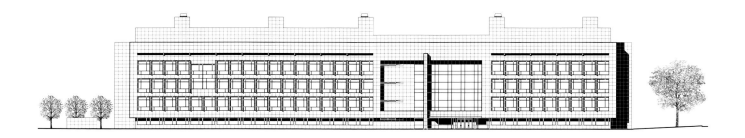

北立面

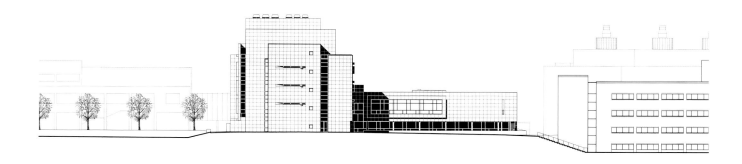

南立面

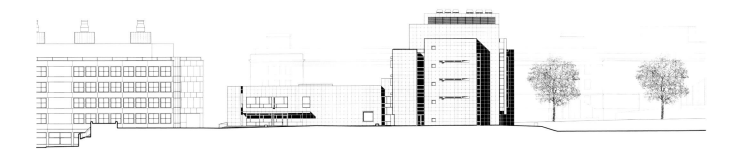

西立面

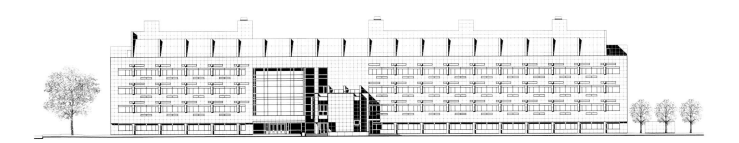

ECM城市大厦
ECM City Tower

ECM城市大厦
ECM City Tower

捷克共和国布拉格,2001—2008

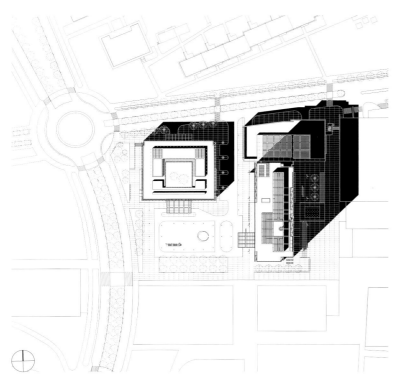

理查德·迈耶及其合伙人事务所为布拉格的ECM城进行了总体规划,城市大厦作为其中的一部分,是布拉格城市的标志性高层建筑。这个设计体现了朝气蓬勃而又不失严谨的形式语言,同时满足了既有结构和扩建结构的技术需求。设计将使用面积、效能和灵活性最大化。玻璃幕墙从地面延续到屋顶,内部可观赏历史名城布拉格无与伦比的美景。

从古老的布拉格城市中心可以看到建筑符号式的西立面,它重新定义了捷克立体主义,立面上连续的板片折叠或展开,创造了一种抽象的、城市尺度的窗,可俯瞰城市中心。那些重要的楼层顶部延伸的窗户提供了更多可使用的空间。玻璃幕墙以1.5米的深度起伏,它模糊了观者对主导面的感知,产生了在前景和后景之间丰富的光影关系。展开的面板用第二层玻璃延伸至建筑的边缘,作为双层表皮来定义内部空间,也展现了高层建筑的尺度和比例。

从远处看,建筑可从城市背景中区分开来。戏剧性的是,行人从近处街道同样可以看到这种效果。立面下部的附加层用于营造街道层面的视觉效果,定义了建筑的底部。建筑的底层有如下特点:入口雨棚具有雕塑般的特征,用餐平台沿着金属板包边的两层大厅布置,另外,两个咬合的体块和南北两面垂直金属墙表达了纤细的形体轮廓,调整了建筑的比例。

对称

结构

入口

流线

围合

垂直/水平

地面层平面

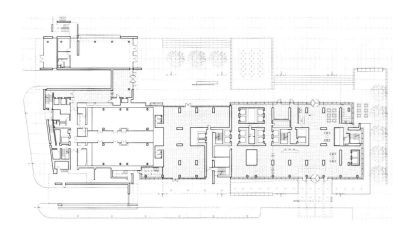

13层平面

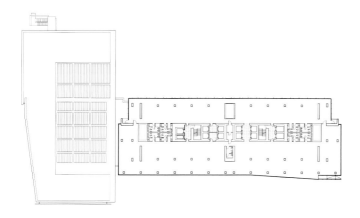

北立面 东立面

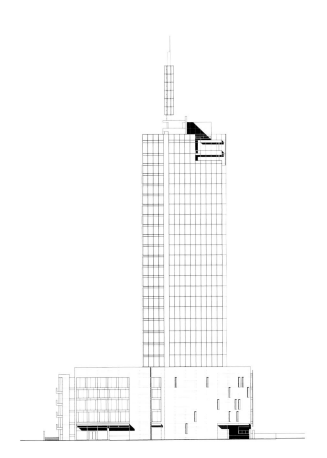
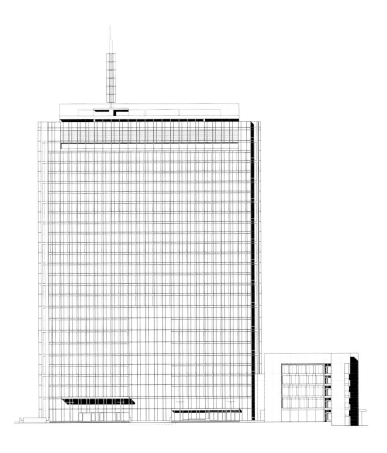

Feldmühleplatz 办公楼
Feldmühleplatz Office Building

Feldmühleplatz办公楼
Feldmühleplatz Office Building

德国杜塞尔多夫，2002—2004

现有Feldmühle办公楼的两个新的加建项目被委托给迈耶事务所，设计目标是为大型国际律师事务所创造现代化的办公环境。由此产生的Feldmühleplatz办公楼包含了46 000平方米的高端办公和相关设施。设计在满足客户要求的同时，创造了一种整合大型办公综合体及其住宅和公园般环境的空间状态。通过每扇窗都可以看到田园般的景色。精选的材料与最先进技术系统的结合让这栋建筑成为21世纪的办公楼。

场地位于杜塞尔多夫，地理位置优越，可观赏莱茵河和市中心的景色。在19世纪末，场地周边环境产生了独具特色的城市肌理，并保存至今。场地较为平坦，北边临近办公建筑和公共公园，南边是住宅区花园，东面是Feldmühleplatz，西面是四层的住宅建筑。新建建筑为五层，与原有的六层建筑相连，原有建筑已经被全面翻修。新的会议中心加建在屋顶，可观赏壮美的景色。屋顶会议中心也是新建筑之间的视觉连接。

新建筑采用了技术创新产品，包括高性能轻钢、带有机械遮阳装置的透明玻璃窗，个人用户可独立地对其进行操作。为了增强私密性，办公室隔墙具有很高的降噪等级，机械系统中含有特殊的隔音消声器可加强降噪效果。机械系统是自然通风、集中机械送风和空调送风的集合系统。可操控的窗户完成自然通风，空调送风在办公室和会议空间均可独立控制。

流线

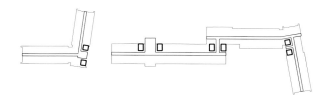

入口

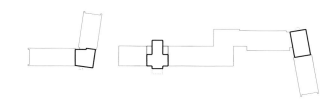

现有和更新

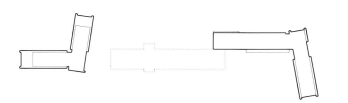

几何

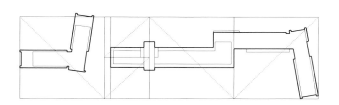

正交和斜线

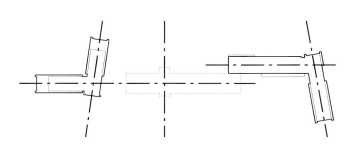

结构

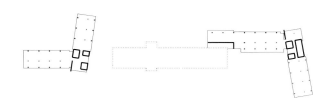

地面层平面　　　　　　　　　　　　标准层平面

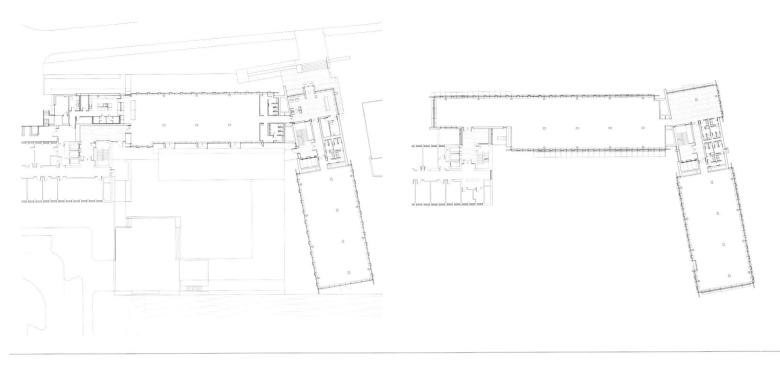

四层平面

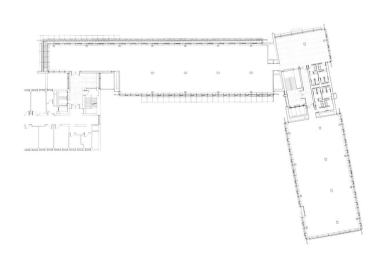

东立面

西立面

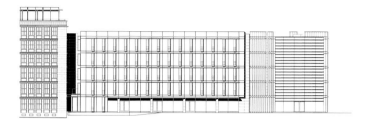
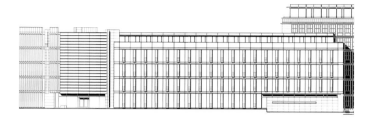

北立面

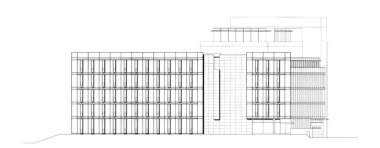

耶索洛丽都村、公寓房和酒店
Jesolo Lido Village, Condominium, and Hotel

耶索洛丽都村、公寓房和酒店
Jesolo Lido Village, Condominium, and Hotel

意大利耶索罗，2003—2010

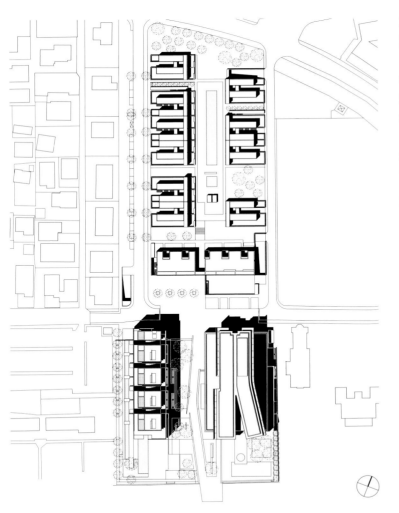

耶索洛丽都项目，坐落在意大利亚得里亚海岸一个标志性的海滨旅游目的地。该项目由2007年完工的耶索洛丽都村、与相邻住宅建筑规模相同的三层住宅综合体和两个海边的建筑（耶索洛丽都公寓和耶索洛丽都酒店）组成。耶索洛丽都村由带有底商的面向中央广场的长方形住宅建筑和两排大小不同的住宅公寓组成，公寓两边是游泳池和公园空间。该村子布局中的典型住宅模块设有公共外部楼梯和同一水平面上的小花园，在三个楼层上，把一居室和两居室两种单元联系起来。这种独立式的住宅模块提供了实与虚的交替，面对游泳池的立面上是连续的百叶窗系统。

耶索洛丽都公寓应用建筑学常见的语汇，开放的入口广场连接村子，使两个项目之间产生对话。地面层包括六个带有私人花园、温泉区和礼宾台的公寓。建筑的顶层被分配给五个不同的顶层复式公寓单元，每个单元都有私人室外游泳池。这个10层建筑，地面以上的面积大约有7400平方米。

耶索洛丽都酒店为建筑组群增加了第三个元素，整个建筑包含两层高的裙房部分，内部是大厅、健身设备、餐馆，裙房以上的四个楼层包括122间客房，每一间都能看到亚得里亚海。

村子
地面层平面

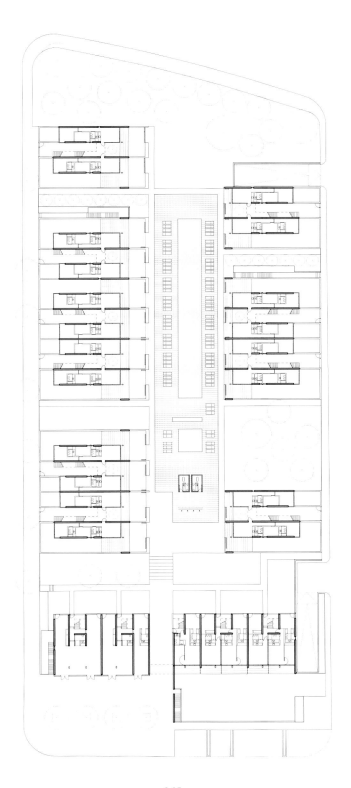

高层住宅
一层平面

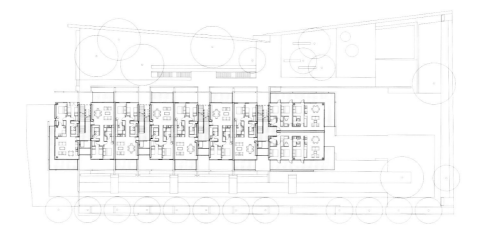

高层住宅
八层平面

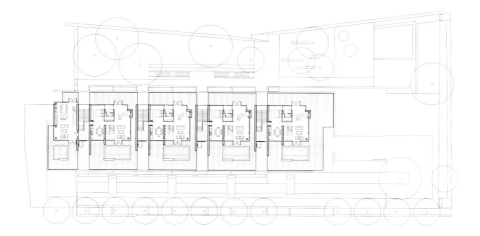

高层住宅
北立面

高层住宅
南立面

高层住宅
东立面

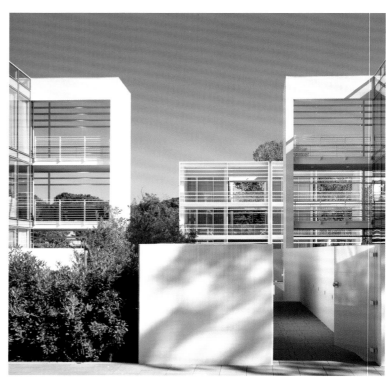

圣丹尼斯办公楼
Saint-Denis Office Development

圣丹尼斯办公楼
Saint-Denis Office Development

法国巴黎，2003—2009

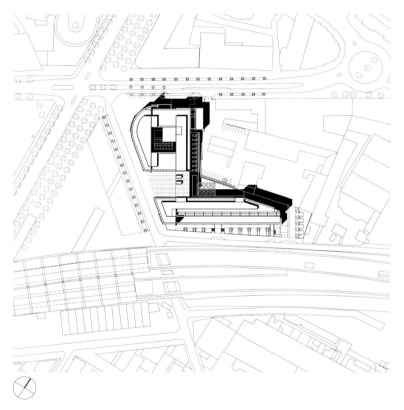

这对新建筑位于巴黎北部圣丹尼斯区，沿着Place des Droits del'Homme的一边布置。Place des Droits del'Homme是一个公共广场，从广场沿去往法兰西体育场的主要道路可直达火车站。项目总面积3.53万平方米，其中包括主要办公空间，也包括厂房、餐厅、自助餐厅和零售空间。两个主要的建筑体块都是地面以上七层，地面以下两层停车场。六层的中庭连接着两个建筑体块，提供了面向广场的优雅入口。

项目的形式回应了周边的文脉、场地边界以及分区和规划的要求。两个建筑体块在一层和七层采用立面缩进的形式。街道层级上，这个项目提供了一个廊道来吸引公众，特别是广场方向的。

项目与Place des Droits de l'Homme的关系体现在它的方向上。从广场看，每个建筑都保持着自己的特征，同时使用了相同的建筑语汇。以这种方式，建筑犹如双胞胎一样，通过一个透明的中庭连接，相似但并不完全相同。

几何

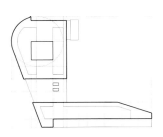

结构

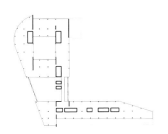

入口

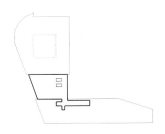

流线

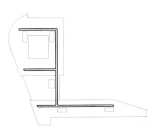

围合

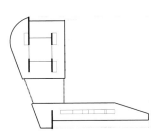

图底关系

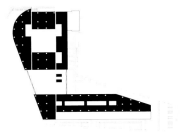

地面层平面

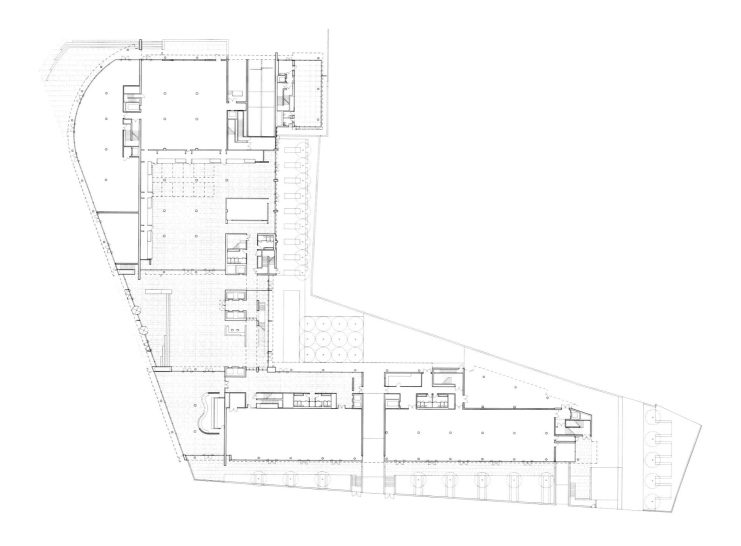

南立面

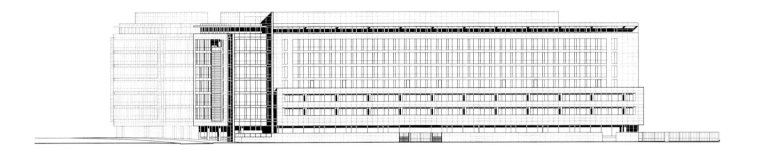

西立面

美国联邦法院
United States Courthouse

美国联邦法院
United States Courthouse

加利福尼亚圣地亚哥,2003—2011

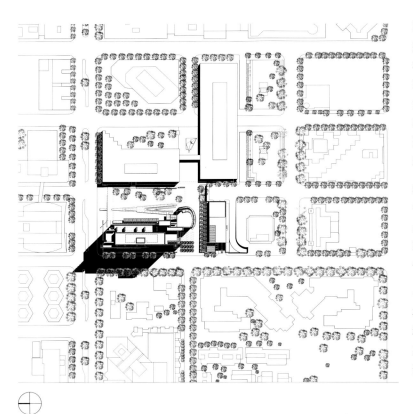

除了新的法院之外,本案的总体规划将花园、广场、水景、人行道和联邦建筑原有部分和新建部分连接起来,以实现中心城区城市设计的目标。为了平衡项目自身需求和公共设施需求,设计将面积达9290平方米的场地创造性地利用起来。地下配套服务区域占据整个场地,而广场层的建筑占地,只有1858平方米。该规划的策略是让场地的中心成为重要的新的公共广场,它将会成为联邦建筑综合体中一个活跃的中心,也是这个城市不可或缺的新的市民活动空间。设计遵循了可持续和能源高效利用的原则,以达到LEED银牌认证的标准。

新联邦法院的建筑体包括修长而优雅的16层高的塔楼,底部是透明和半透明的建筑裙房。塔楼外表是非常薄的陶瓦和玻璃覆层,它们按照对应建筑空间和朝向的原则进行排布。超薄的体块体现了可持续设计的策略,(这样的设计)用来为整座建筑提供自然光,更重要的是为法庭提供自然光,法庭里可以从东、西两侧接收并过滤阳光。

与直线型高层部分并列的建筑大厅是一个标志性的椭圆形体块,该体块的位置经过严格推敲,保证衔接场地的道路。大厅的形式用于接纳和指引员工及来访者去往不同的法院。大厅中充满了从玻璃屋顶洒下来的阳光,多种形式的夹层活跃了大厅空间,同时也完善了员工和来访者的流线。

入口

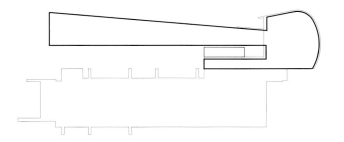

流线

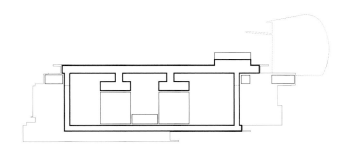

围合

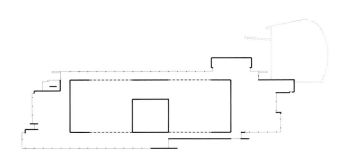

图解

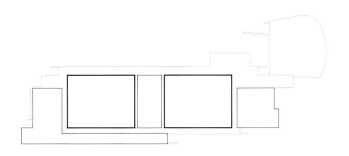

法院入口层平面

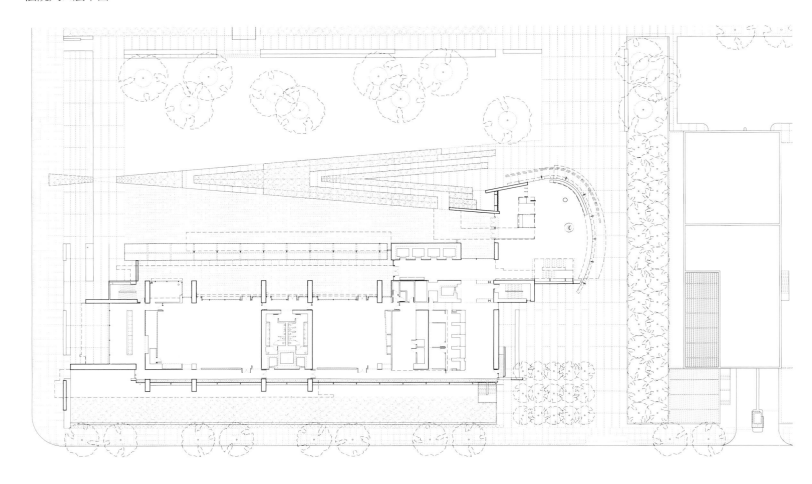

法院标准层平面

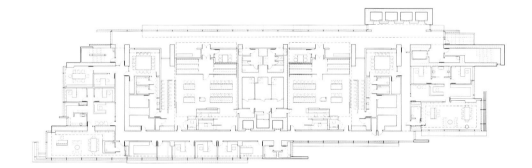

东立面

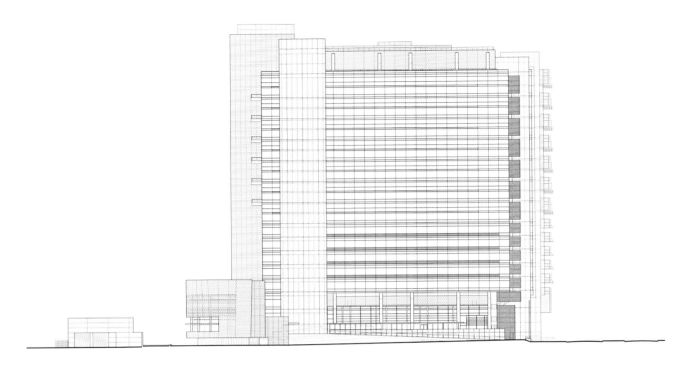

北立面

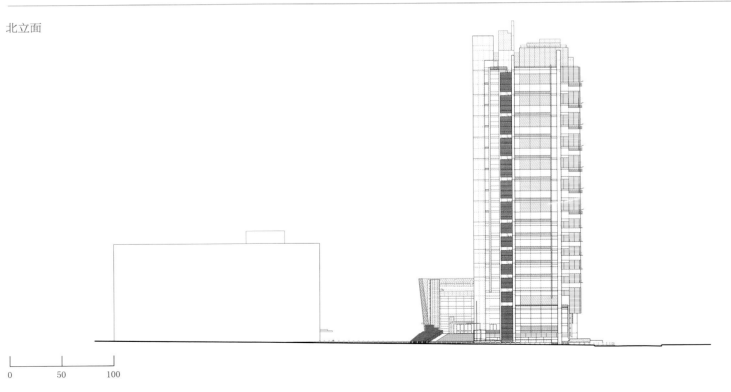

0　　50　　100

南立面

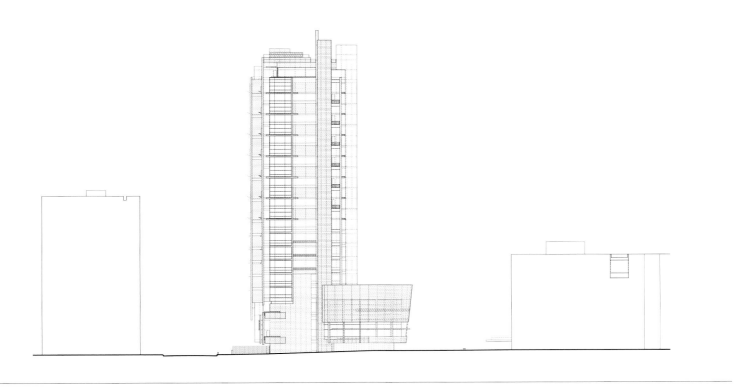

西立面

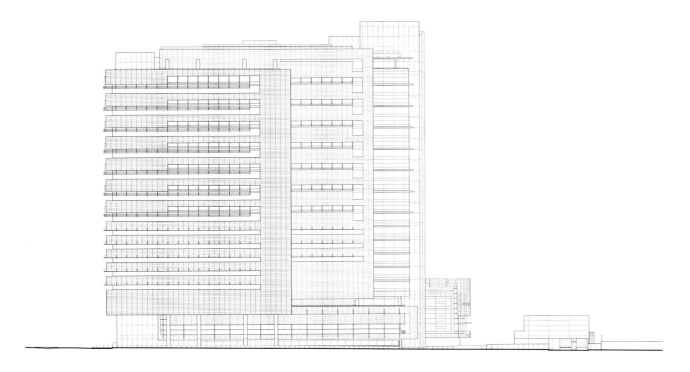

东江总体规划
East River Master Plan

纽约，2004—2012

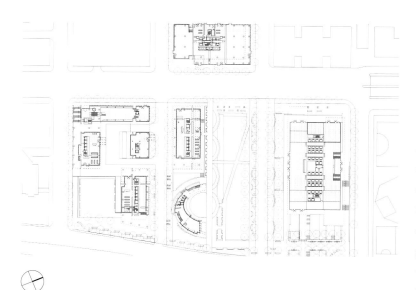

东江总体规划将开发纽约市临罗斯福路的四块场地，位于联合国大厦以南，从东四十一路到东三十五路的区域。该项目中的四个高层住宅由理查德·迈耶及其合伙人事务所设计，其中之一位于第一大道的西侧，另一个商业综合体大楼由SOM设计，坐落在三十八和四十一大道之间，第一大道的东侧。

该项目约有75％的土地位于城市开放空间中，项目必须为一个大型城市公园留出空地，公园内有植物、座位区、溜冰场和其他设施。我们希望沿着第一大道设计遍植树木的林荫大道，从第一大道到东江的人行道（坐落在三十九和四十大道上）也一样将会是枝叶繁茂的绿树景观。人行道边是精心设计的公园景观，将会把周边建筑都连接到可以俯瞰水景的人行道上。

整个项目的场地允许的最高容积率为10，所以设计中的建筑的高度都在148米和185米之间。规划建设面积约为44万平方米，包括约2500户新建住宅。公共停车场和住宅配套停车场位于地下。新的主要零售空间面向第一大道，迈耶设计的餐厅将会毗邻人行道，可观赏东江的景色。

几何

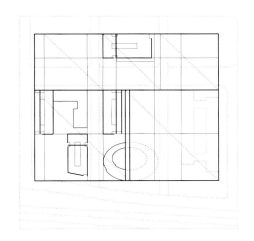

景观

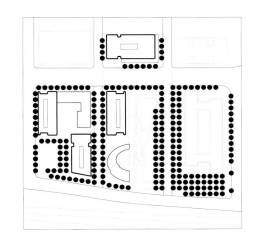

围合

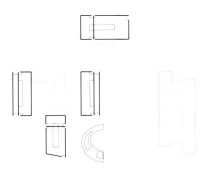

结构

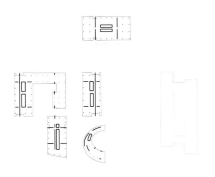

入口

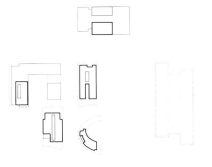

流线

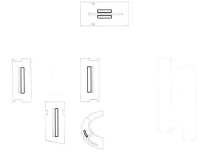

地面层平面

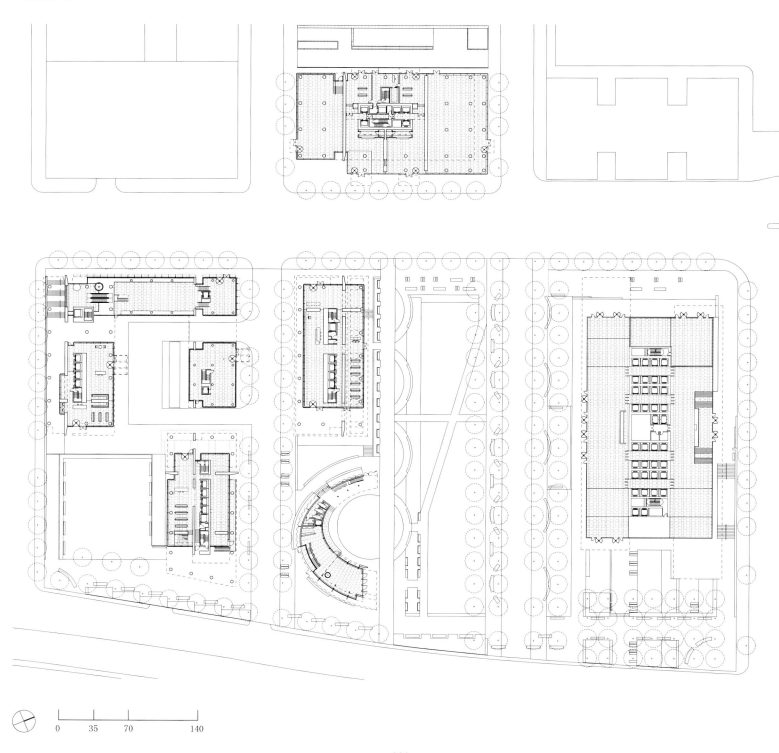

东立面

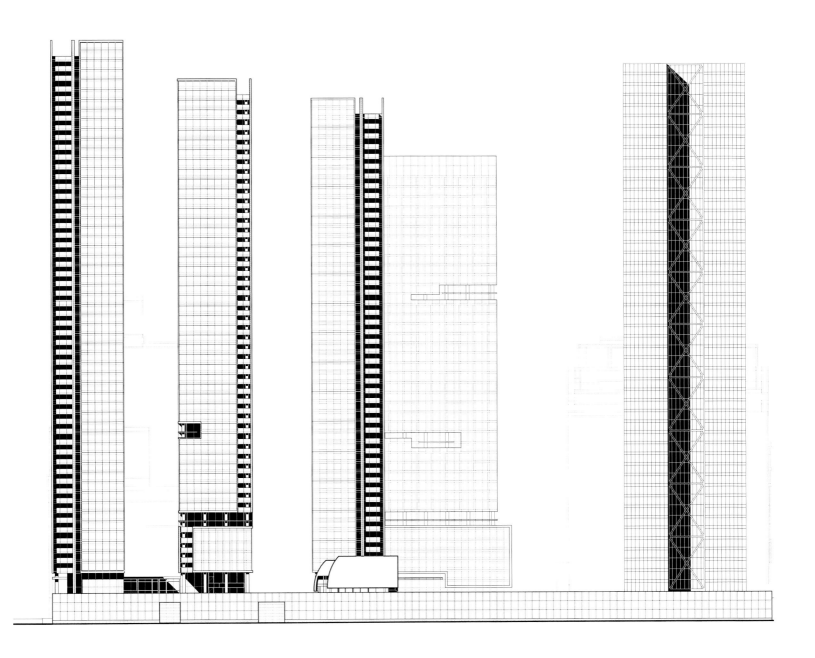

北立面

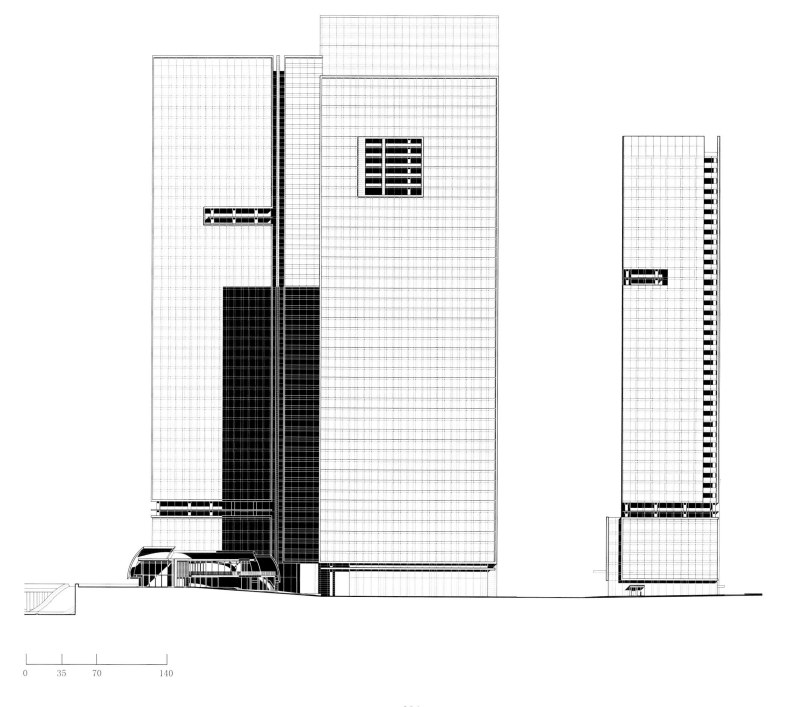

南立面

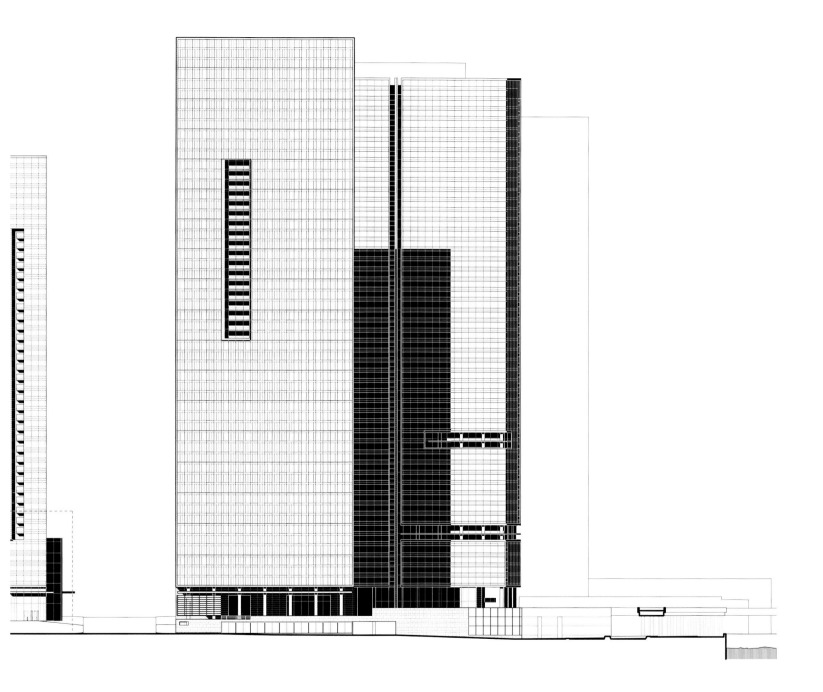

咖啡广场
Coffee Plaza

德国汉堡，2004—2009

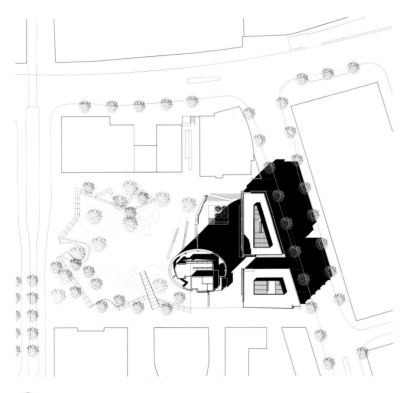

坐落于汉堡港口新城的咖啡广场是为国际咖啡贸易和相关商业业务而规划的一个特殊区域。该项目涵盖了全球最大的咖啡贸易公司卡夫总部、两栋可出租的办公建筑、公共广场和地下停车场。本案坐落在Sandtorpark旁，紧邻可俯瞰港口新城西部Sandtorhafen的麦哲伦平台，是集城市生活、休闲游憩和商业活动功能于一体的城市枢纽。

建筑坐落在一个可俯瞰Sandtorpark的平台上，椭圆形的高层建筑将公园和广场分隔开来。建筑面积共27 100平方米，12层，地面以上11层是办公空间，附加的顶层阁楼是会议室。高层建筑还包括三层地下层，为停车场和设备用房。

两个附加的结构合体，形成了广场的取景框。每栋建筑均为地下一层、地上六层，同时还有屋顶露台以及两个地下层以容纳停车场和设备房。地面层均用作零售空间。三个建筑都采用了灵活的平面布置。

将建筑和城市肌理的关系作为重要的设计内容进行处理。综合体和广场形成大面积城市开放空间的雕塑般的边界，三个建筑从抬升的广场和划定的城市街区升起，为这片区域提供场所感和社区归属感。

几何	结构
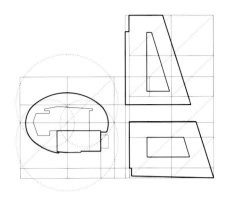	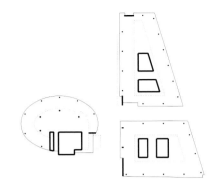

入口	流线
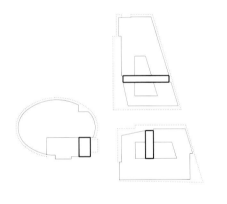	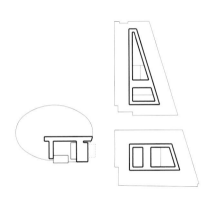

围合	图底关系
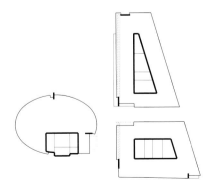	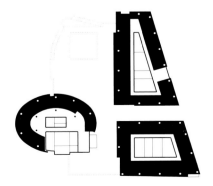

一层平面

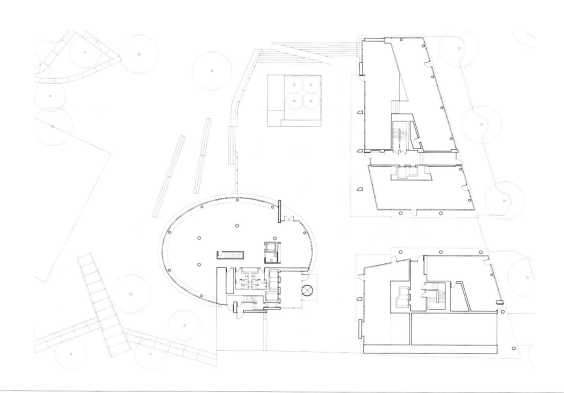

二层平面

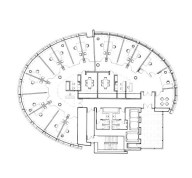

0 2 5 10

西立面

北立面

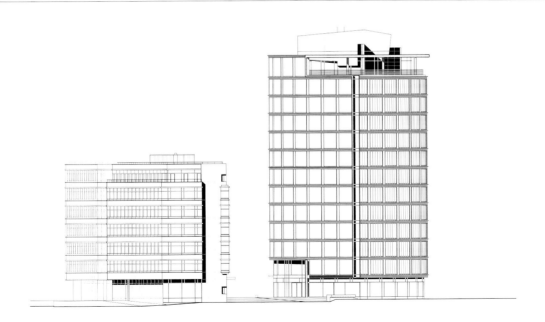

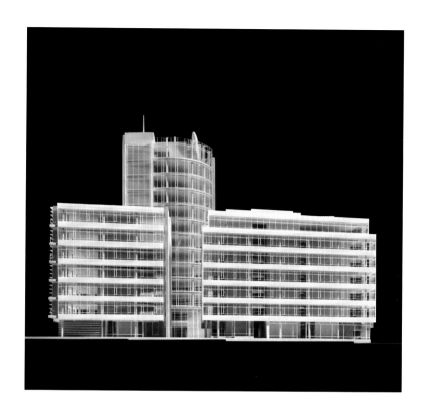

9900威尔希尔
9900 Wilshire

加利福尼亚比弗利山，2004—2011

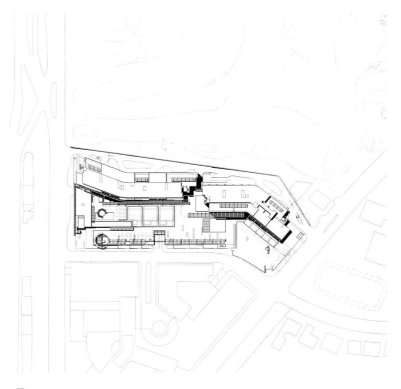

9900威尔希尔由200多个豪华公寓住宅组成，项目由13层和15层的住宅建筑、公共和私人花园、水景和零售商店组成。建筑设计针对特定场地，有一种永恒的存在感，与邻近的希尔顿酒店协调，与邻近的洛杉矶乡村俱乐部呼应。9900威尔希尔体现了标志性场地的重要性，同时也反映了国际形象和"花园城市"的形象。

两栋住宅建筑都位于产权线（用地红线）的西侧，从地面升起以获得最多的自然光。建筑以曲线和水平的形态体现自然的形式，反映乡村俱乐部景观的几何形态，也与邻近希尔顿酒店经典的水平体块协调。许多住宅都设计了通高空间，保证住宅与乡村俱乐部和比弗利山视线连通。设计运用庭院、露台、阳台和大面积的遮阳玻璃，延伸内部空间到外部空间，这些措施适应了地中海气候的特征，使其成为南加州的最佳住宅。

9900威尔希尔旨在通过对环境敏感的建筑系统获得LEED金牌认证。建筑运用可持续发展的设计理念，使用了超薄结构，最大限度地利用自然光和自然风。每间公寓均包括悬臂式露台和阳台，提供遮阳装置的同时在建筑立面上创造了深度感和尺度感。除了住宅和花园，在圣莫妮卡大道上，1394平方米的零售空间将成为比弗利山和世纪城之间重要的连接体。比弗利山的场地中，设计的目标是通过建造捕捉加州生活的本质的建筑，提供高品质的光和空间。

围合

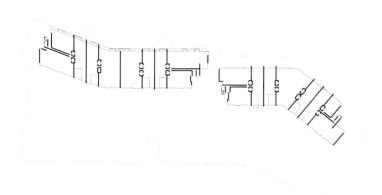

结构

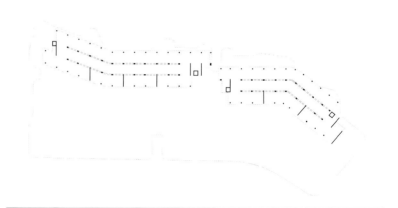

入口

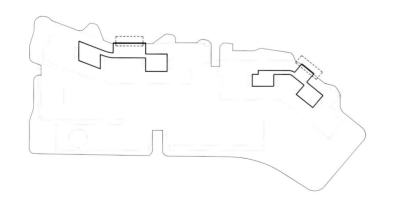

图解

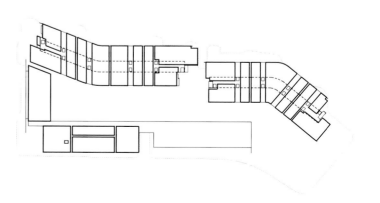

地面层平面

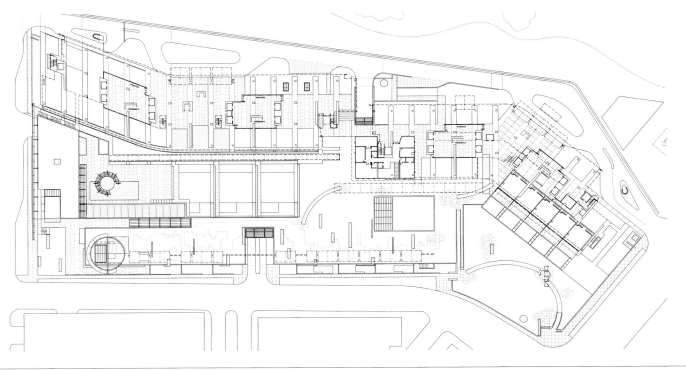

标准层平面

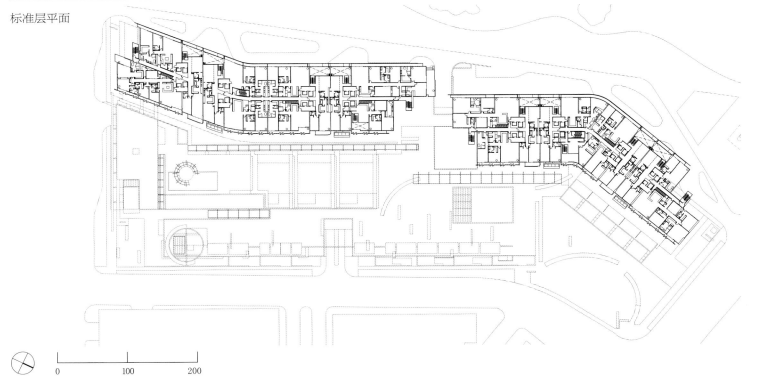

东立面

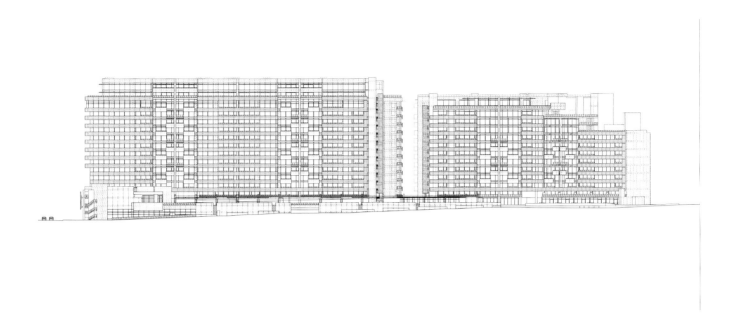

西立面

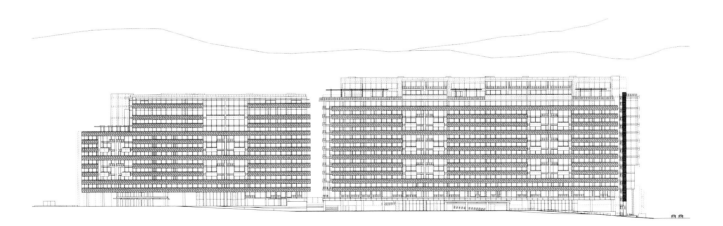

意大利水泥创新技术中心实验室
Italcementi Innovation and Technology Central Laboratory

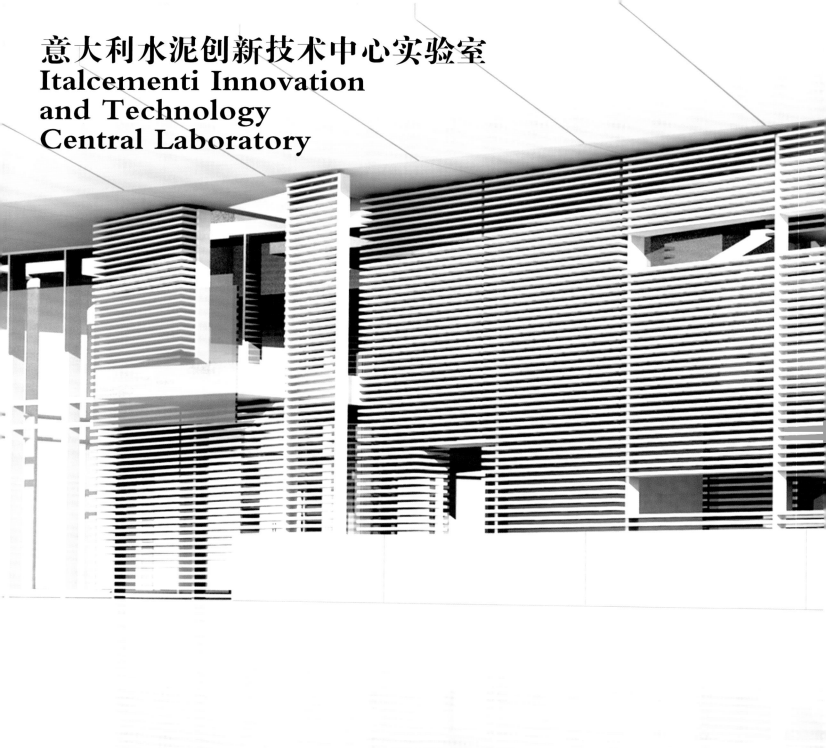

意大利水泥创新技术中心实验室
Italcementi Innovation and Technology Central Laboratory

意大利贝加莫，2005—2010

创新技术中心实验室醒目地伫立在意大利贝加莫的克罗密特·罗所科技公园东端，将成为标志性建筑，表达意大利水泥集团的主导地位，实验室投身于混凝土利用的研究和创新多年，具有被广泛赞誉的技术专长。本项目是可持续设计的标杆，力图成为欧洲第一个达到LEED白金认证的建筑。

新大楼强化了三角形场地的边界，将技术空间和行政空间布置在大楼两翼。地上两层，地下两层（包括停车场）。V形结构把实验室和行政侧翼分开，两侧翼围绕着一个中心庭院花园，该花园作为配套服务空间，可通往地下停车场。庭院低于地面，使地下实验室可以获得自然采光，机房和车库设计采用自然通风。

建筑的东北角，也是科学公园的边界转角，大型的公共广场形成了通往两层带有天窗中庭的主要入口。中央中庭设有接待处和安全控制处，两部玻璃电梯和醒目的坡道为结构的两翼提供交通空间，交通空间均为天窗采光，可以通过中庭两层高的玻璃幕墙欣赏贝加莫的美景。创新技术中心实验室的屋顶为混凝土悬臂结构，仿佛飘浮在空中，形象刚劲而优雅。除了在公共空间采用大型天窗，利用屋顶引导自然光进入办公室、走廊和实验空间外，形成的天窗系统同时也构成了虚拟的第五立面。自然光通过顶层楼板进入室内空间提供照明，光影随时间变化，自然的光与影在与众不同的幕墙上相互辉映，幕墙由意大利水泥集团为此项目研发的混凝土元素、可视熔块和透光玻璃组成。

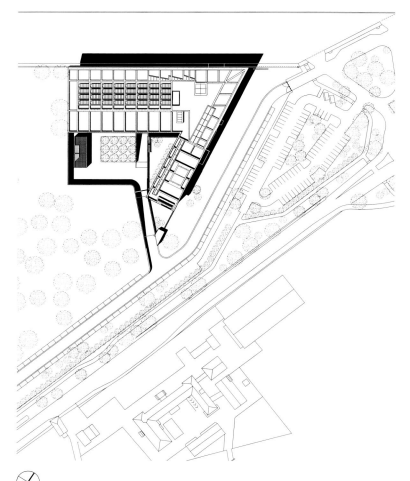

几何

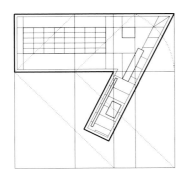

结构

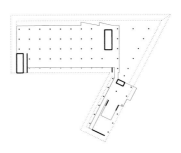

入口

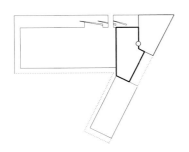

流线

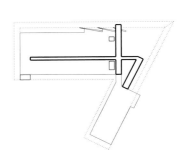

围合

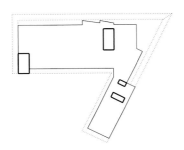

图解

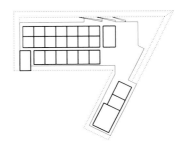

地面层平面

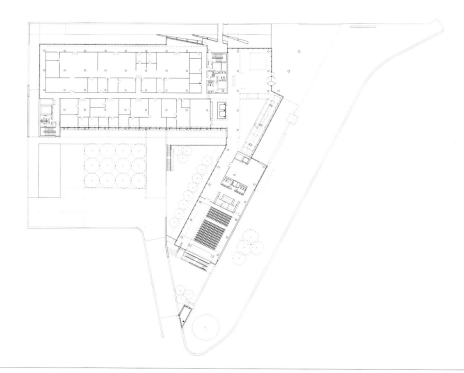

二层平面

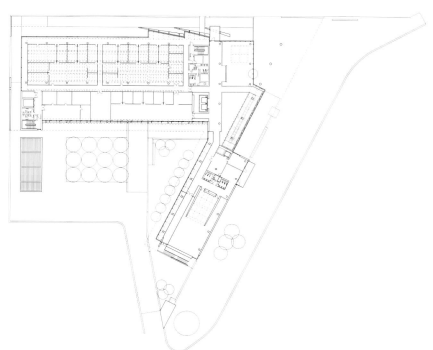

东立面

北立面

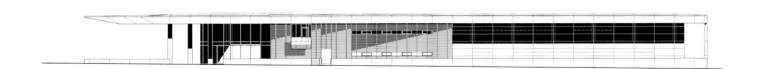

福斯特曼·里特尔公司和国际管理集团世界总部
Forstmann Little and Co. and IMG Worldwide Headquarters

福斯特曼·里特尔公司和国际管理集团世界总部
Forstmann Little and Co. and IMG Worldwide Headquarters

纽约，2007—2008

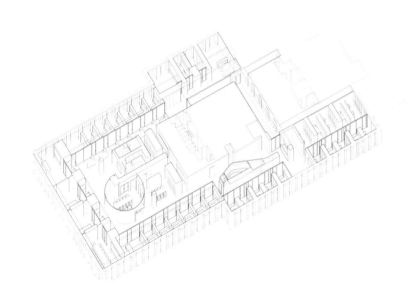

项目位于纽约第五大道的标志性建筑——通用汽车大厦的第45层，容纳了国际管理集团、全球体育、娱乐和媒体公司的行政办公室和私人股份公司福斯特曼·里特尔公司的行政办公室。此为建筑改造项目，原建筑面积1672平方米的空间，其中60%的面积供国际管理公司使用，40%的面积供福斯特曼·里特尔公司使用。两家公司共享公共办公空间，但入口分开，分别位于电梯两端。

董事会室和行政办公室为线性布局，均可获得曼哈顿市中心和中央公园的广阔视野。会议和服务空间，包括浴室、厨房、复印室和储藏室，占据楼层的中心，毗邻核心筒。这种布局产生了一种分层的空间，让自然光通过酸蚀雕刻玻璃进入内部空间，酸蚀玻璃用于分隔办公室与公共空间。在办公室之间使用透明玻璃隔板隔音，保证了私密性，同时改变了传统办公空间的格局。公共空间由两种形式加以强调。中心半透明的圆柱形会议室提供了空间的"壁炉"（空间核心），曲线型的会议室是衔接接待区和私人办公区的过渡空间。

材料以暖色为主。办公室、会议室和董事会房间使用了定制的柚木抛光材质和棕色的皮革家具。大厅和浴室中使用了塞纳石灰石作为饰面材料。为与国际管理中心世界顶级媒体公司的身份相匹配，从接待区大面积的媒体墙到会议室的通信系统和遍及整个楼层的电动窗帘，新技术被运用到每一个功能空间。

楼层平面

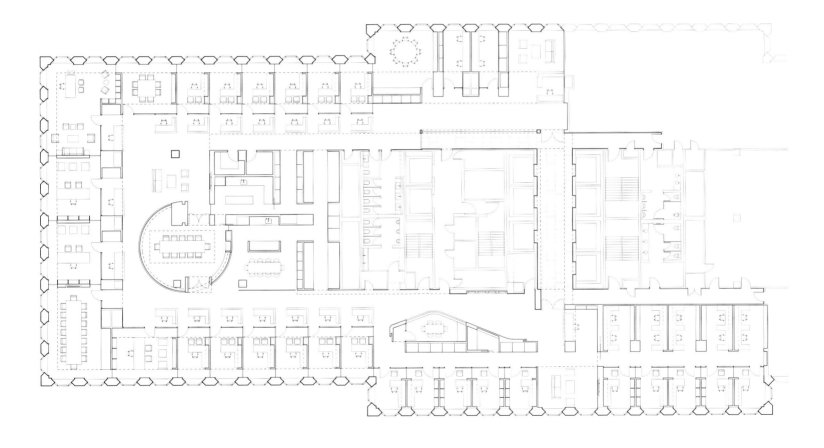

罗斯柴尔德大厦
Rothschild Tower

以色列特拉维夫市，2007—2010

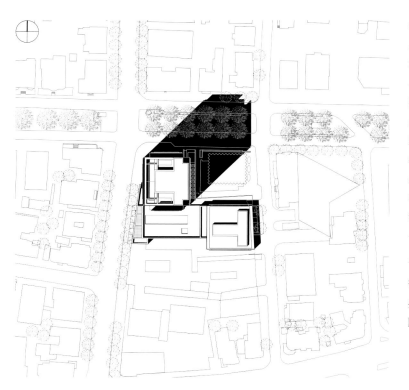

纯净的、具有标志性的39层建筑坐落于罗斯柴尔德大道和爱伦比街，那里是特拉维夫市最著名的地点之一。建筑综合了多种功能，包括住宅、零售和办公设施。大厦设计所考虑的基本因素是光的品质、观海视线和与现存罗柴尔德大道建筑的关系。设计的目标是将建筑固定在场地上，整合特拉维夫市核心区域并成为新的地标，衬托周边的建筑环境，周边多为白色城市历史时期的包豪斯风格的老建筑。大厦的体块简洁而优雅，注重材质的轻快、精美和通透。裙房空间开敞、自然通风，容纳温馨的大堂和零售空间。大厦前广场与街道和步行路近距离接触，在罗斯柴尔德大道边植树以将公共区域与交通区域分开。沿地面层街道立面设有轻型玻璃雨棚结构，二层立面向爱伦比街大面积设有开口，内部设有游泳池，在亚夫涅街一侧将会建造水疗中心，这一切都会给特拉维夫这一著名区域增添新的活力。住宅楼还设有地下酒窖和休息室。

设置零售空间以更新已有建筑的拱廊设计。用于服务爱伦比街和亚夫涅街的走廊入口突出，在亚夫涅街的入口也可到达楼上面积465平方米的办公建筑。建筑的动态组合有助于特拉维夫市作为一个充满生气、带着欧洲都市精神的城市的持续发展。

几何

结构

围合

入口

地面层平面

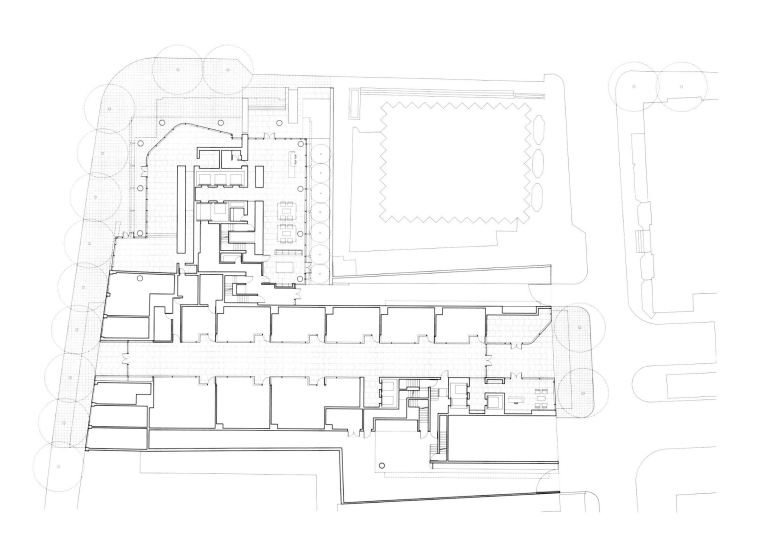

二层平面

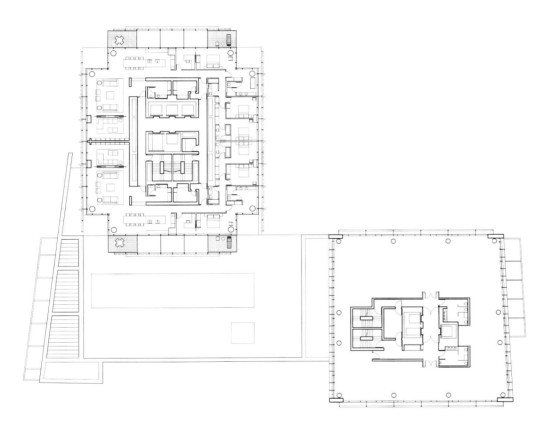

北立面 东立面

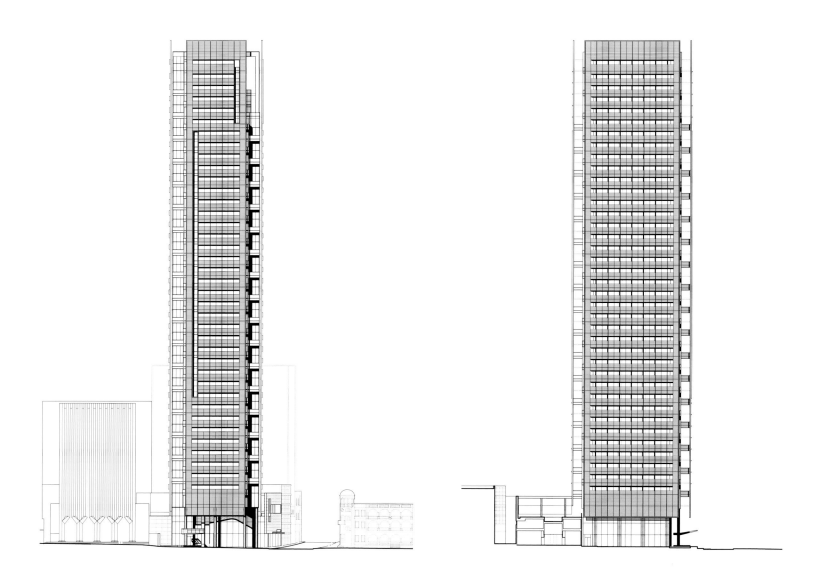

0 5 10

南立面 西立面

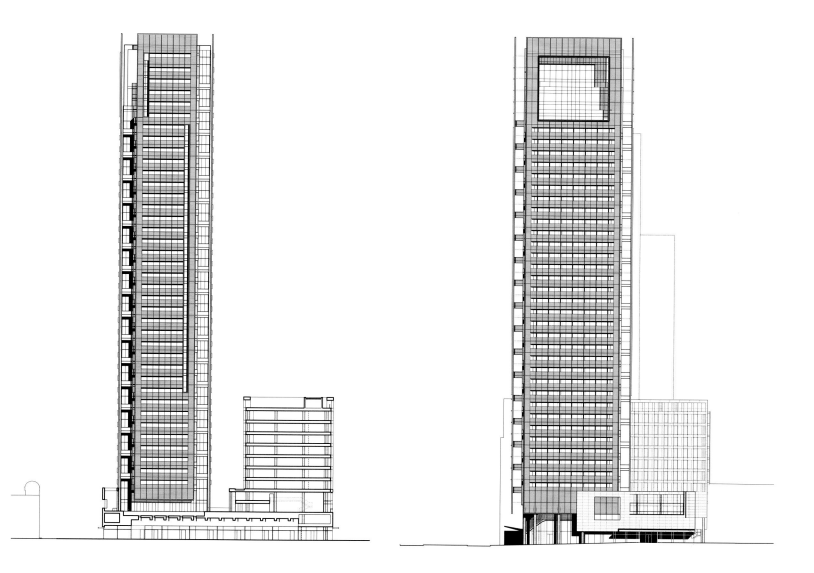

纽瓦克苏玛总体规划
SoMa Newark Master Plan

纽瓦克苏玛总体规划
SoMa Newark Master Plan

新泽西纽瓦克，2007—

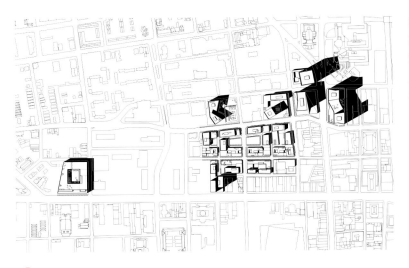

纽瓦克苏玛总体规划为纽瓦克市中心马凯特街南部区域（这片区域称为苏玛）的成长和蜕变提供了一个综合远景。由此产生的多种功能、尺度和密度都会注入该区域，设计用充满活力的空间状态表达社区的当前需求和未来潜力，本案是全美城市规划和城市发展的典范。总体规划由四部分组成：在规划的核心位置，低层和中高层建筑组团围绕着中心公共空间定义了街道景观。整个项目的灵魂在于强调人体尺度和各种社交公共活动。规划还包括：沿华盛顿和布罗德街布局的中型塔楼，它们与宽阔的车行路和人行道的尺度相匹配；一组面向城市最繁忙的商业走廊之————马凯特街的高层，将构成纽瓦克市的未来天际线；西侧的组成部分，灵感来源于芝加哥林肯公园，以其街道前各种各样的城市尺度的空间营造著称。

该项目策划之初设定的目标为纽瓦克市21世纪完成的第一个大尺度、具有里程碑式意义的建筑。混合用途的建筑被设计成为两个高大而修长的77层塔楼，建筑的裙房占据整个场地。双子塔形体相同、形象鲜明，因其位置临近而不断地进行对话。两个塔楼位于场地对角，在二者衔接的过程中，创造了一个起伏的中央部分，沿着马凯特街和华盛顿街不同的角度展现了不同的形象。架高的部分形成建筑中庭，也创造了一个城市窗口。因为立面表皮包裹和环绕了整个建筑，塔楼可以被看作是裙房的一部分，是地面上的形态，或者被看作是与裙房连接的裙房延伸物。连接和定义建筑、空间组织和结构的不同部分形成了一个整体结构：大胆和戏剧化的轮廓可从所有重要观景点包括曼哈顿清楚看到。

体块/开放空间	街道层级
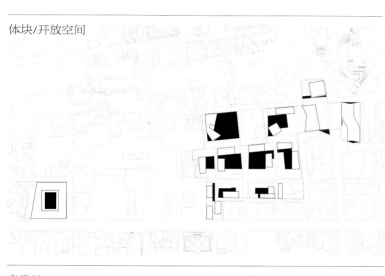	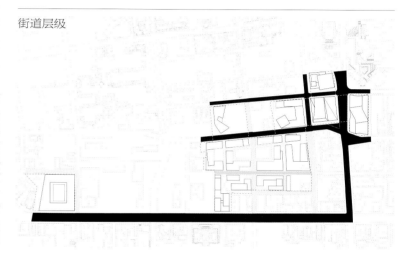
多孔性	绿色网络
	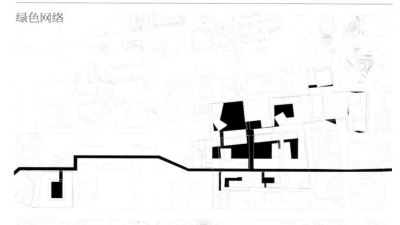

地面层平面

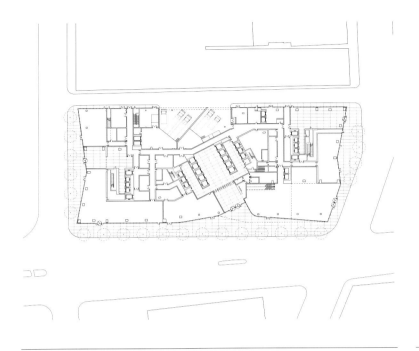

六层平面——空中大厅

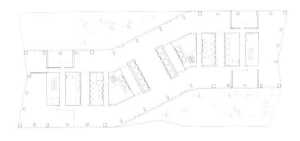

标准办公层平面

典型住宅层平面

0　40　80

马凯特街立面

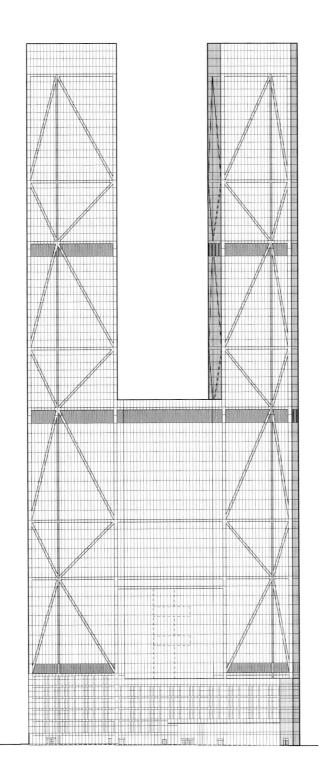

天津酒店
Tianjin Hotel

中国天津，2009—2012

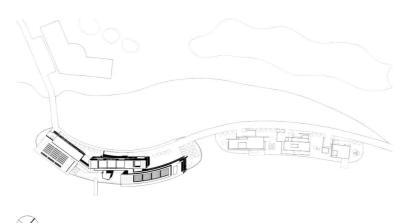

在一片岛屿场地上的8星奢华酒店在被称作"世界"的天津星空传媒集团的开发区内（星耀五洲项目），它提供31个客房套间，还有最先进的水疗中心和健身中心。客房套间在两个互补的体块里，东侧翼和西侧翼，两个侧翼创造了沿岛屿伸展的动态形式。一个直通建筑顶端、18米高、顶部采光的中庭连接两个体块，创造了一种光与影的动态互动关系。酒店包括餐厅、私人用餐套间、酒吧、图书馆、商业中心核、屋顶露台等，除此之外，还有一个俯瞰花园和海滨的画廊以供活动使用。

客房套间面积很大，内设蒸汽浴室、桑拿设备和可欣赏外部景色的浴缸。东侧翼的第五层是两个高级套房，每一间都配有两间卧室、起居室、用餐区和服务厨房。

水疗和健身中心坐落在岛屿的西端，以一个廊桥连接客房套间。那里包括治疗室、休息室和带有25米长顶部天窗的小型游泳池。这个建筑质朴，为使其几何造型与酒店侧翼协调，也由两个体块组成，形成一个整体，虚实相间，从东至西形成连续的空间序列。

地面层平面

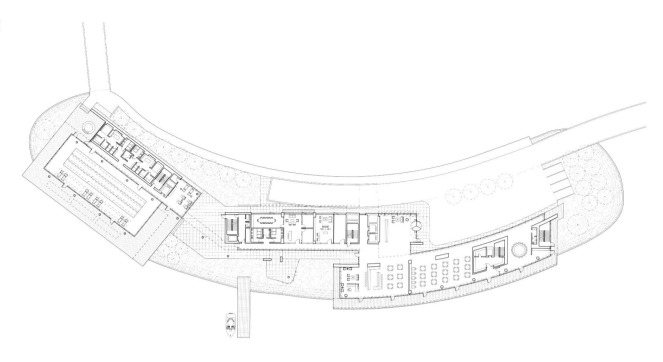

二层平面

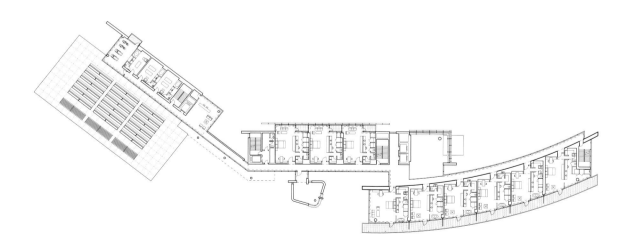

南立面

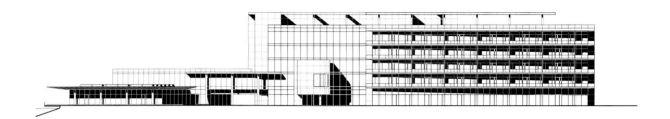

北立面

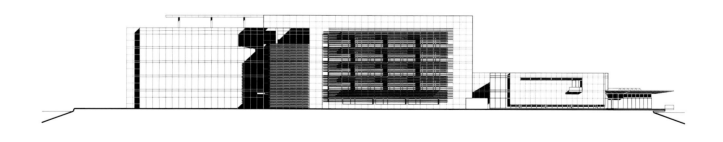

0　　10　　20

大厅横剖面

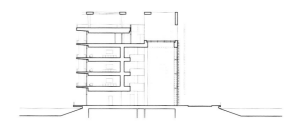

画廊和酒店套房横剖面

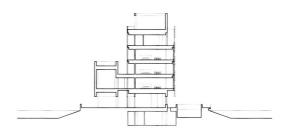

贝多芬音乐节剧院
Beethoven Festspielhaus

贝多芬音乐节剧院
Beethoven Festspielhaus

德国波恩，2009

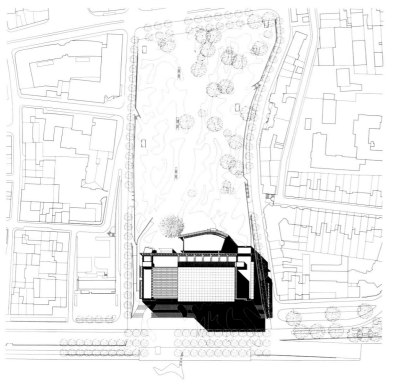

这个在波恩（路德维希·范·贝多芬出生地）为音乐厅和演出排练设计的空间有一个构想，那就是在莱茵河河岸美景中以室内外空间连接城市音乐遗产的计划。本案提出了贝多芬高地的城市设计概念，用大型城市公园连接在博物馆大道的文化机构和河流，由迈耶与景观建筑师詹姆斯·科纳的菲尔德景观设计事务所合作设计。

设计围绕三个主要空间进行组织：音乐厅、排练厅和公共休息室。音乐厅横跨滨河步道，从莱茵河方向即可看到演出场地，同时，从公园方向独奏厅清晰可见。门厅连接两个大厅的公共入口，醒目地表达了从公园到河流的空间过渡。过渡空间中是一系列为主要空间配套的多功能室。

位于建筑核心的中庭空间吸引观者从主要入口进入，走向东立面大面积的玻璃幕墙，以欣赏莱茵河美景。中庭下方在临河一侧设贵宾停车区入口。中庭设计的主要概念是成为公园和莱茵河滨河步道的连接体。大型公共楼梯创造了在水平方向的过渡，引导观者通过建筑从公园到达河岸，反之亦然。音乐厅的外部是一个压花铝板的"盒子"，让自然光经过铝板的空隙进入音乐厅。盒子表皮在白天呈现强烈的、立体的形式，晚上则会是漫射的光幕。

地面层平面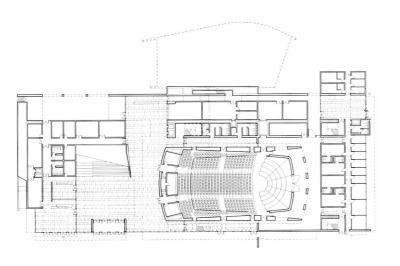

二层平面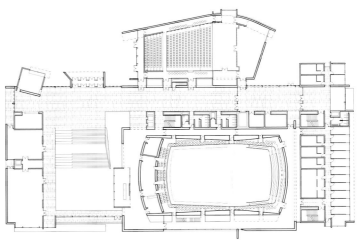

三层平面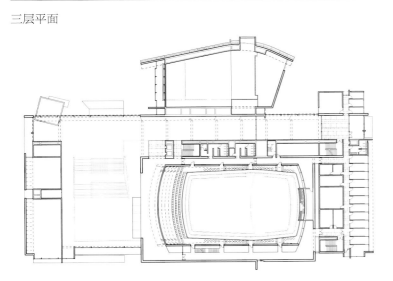

四层平面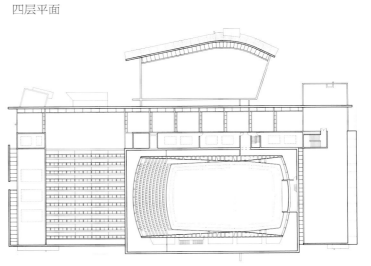

东立面

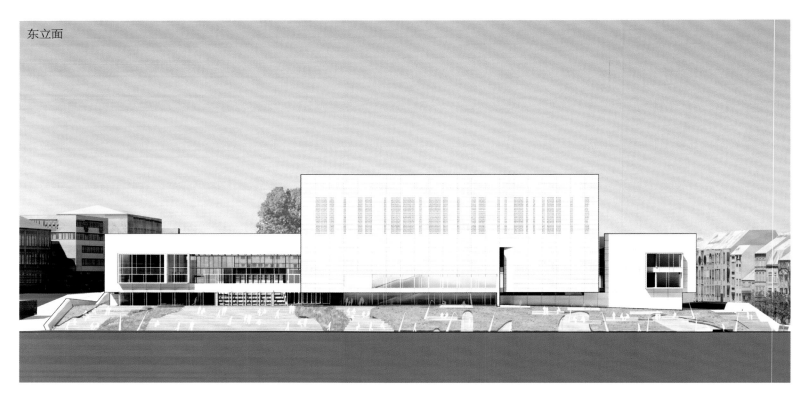

北立面

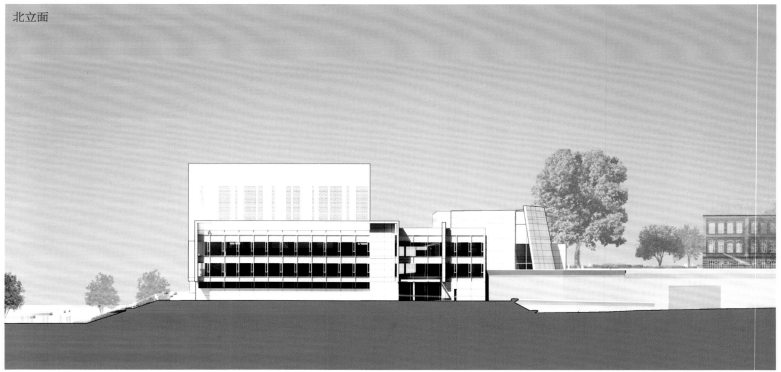

南立面

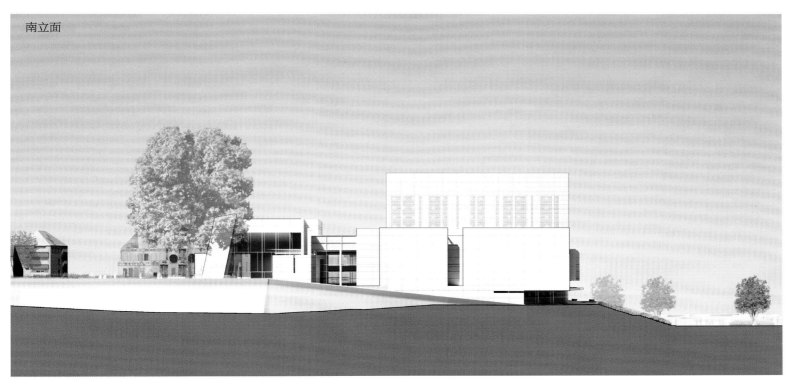

西立面

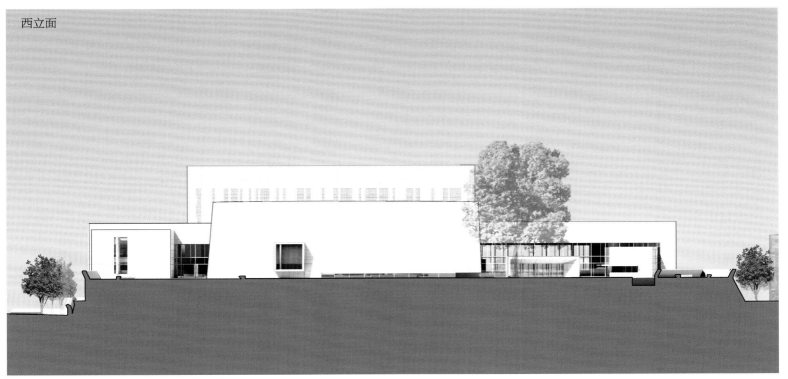

编后记
附言
理查德·迈耶简介
年代表
理查德·迈耶及其合伙人
事务所模型博物馆
参考文献
合作者
顾问
图片注解

编后记
Afterword

保罗·戈登伯格

关于理查德·迈耶的建筑，还有什么未曾表述？写迈耶第五卷作品集的编后记，从某种程度上来说，需要一个定论，一个法学学者最喜爱的最伟大的拉丁短语总浮现在我的脑海里，res ipsa loquitu，这句短语一般被翻译为"事物为自己说明"，在最后，简单地让迈耶的建筑去为他代言吧。但是res ipsa loquitur的直译与普通法学词汇的意思不尽相同：它不是"一个事物为自己说明"，而是"事物本身会说话"，当你思考建筑的时候，词语的微妙变化让事情变得完全不同。这并不表明解说是多余的，而是建筑本身有能力跟我们对话，我们有义务倾听。你可能说，我们必须给迈耶的建筑说第一句话的权力，在那之后，该轮到我们来表达了。

当然迈耶的建筑总是具有这样的话语权；没人会看到一座迈耶的建筑而迷惑建筑师究竟在思考什么，或者感觉在他开始理解这座建筑的关键点之前需要解释说明。迈耶的建筑总是出于本能进行交流，即便是在表达复杂想法的时候。迈耶的建筑在它是任何事物之前，首先它是一个感性的体验。它是对实体和虚空、光和纹理的探索，也是一篇组织有序的文章。迈耶的每一座建筑都具有视觉冲击力。几乎无一例外，它们都有美妙的形式。

经过45年的实践，迈耶既是相同的又是不同的。说他是相同的，是因为他的作品持续地反映着一种渴望，从20世纪20年代以来持续发展欧洲现代主义建筑语汇，使之成为比早期现代主义建筑更感性和更独特的设计语言，技术上也更为先进。多年的实践证明，迈耶的建筑设计语汇尽管来源于机械时代，它更是数字时代的适宜的表达，这是因为它纯净、质朴的品质比后工业化时期建筑少了些许工业化的味道。带有电路板般的精致感，而非机器时代建筑的笨拙。迈耶使用这样的建筑语汇可能已经超过40年，但是对数字时代及其美学特征的敏锐感知，反而让他的高度个人化版本的纯粹主义建筑在此时此刻——比他刚开始从业的时候——成为更适合时代的表达。

迈耶并不追随潮流。在45年内的实践中，建筑超越时尚，力图展开关于美学的探讨，随着一个又一个建筑作品的出现，很显然，这场讨论是旷日持久的。美学讨论的基础一直没有改变，但是它的许多细节已经发生了转变。迈耶的早期建筑，包括一些标志性建筑，比如史密斯住宅，它们都趋向于平整的表面；在史密斯住宅里，精彩的玻璃立面被当作二维元素阅读，那是迈耶第一个真正的抒情作品。建筑本身当然是三维物体，显然它也围合了空间，但是围合空间的墙本身是二维的。在20世纪70年代、80年代和90年代，从某种程度上说迈耶的作品可以一起被描述。这些作品，表达了从二维墙体向三维空间发展的进程，通过漂浮的楼板、金属百叶窗和其他元素表达材质纹理，这些元素让他的建筑远比之前的丰富和复杂。社区文化馆的设计标志着这个过程的开始，这一时期的作品表面上是复杂的，看起来是随意的，而且非常依赖拼贴的做法。在设计艾斯利普和菲尼克斯法院大楼时期，以及之后的佩里和查斯街公寓时期，迈耶正在创造真正的大尺度作品，而非通过拼贴达到大尺度，在其从业生涯中这是一个重要的转变。自那以来，迈耶已经有不少其他大型建筑作品，例如迈耶母校康奈尔大学的威尔大厅，这个作品进一步证明迈耶在大尺度作品上的造诣。威尔大厅的案例里，立面在运动场旁边，作为明确的、起决定作用的墙是高度活跃的三维空间，换句话说，形成了平坦和高低起伏的合理组合。威尔大厅还揭示了迈耶自采用了立面的视觉韵律以来，从可被称作为随意的形式操作上到底走了多远。甚至连剖面中的一个窗子，另一个剖面中高起的玻璃墙——所有都是内部有意安排的元素，例如教室、实验室和入口大厅。

经过这些年，迈耶也开始自在地面对形状，特别对圆形不再那样执着了。格罗塔住宅中抬升的半圆形起居室是迈耶第一次充分实现这种类型的形式，这个空间讲述了一种带着低调优雅的自然，容易让人回想起密斯；从那之后，他无数次回到这种形式，从乌尔姆到比弗利山，再到圣何塞，此外还有圆形的、体块的概念，这些概念对迈耶来讲是很自然就可以达到实体和虚体的优雅平衡的，这种平衡第一次以独具匠心的方式展示着他的设计。偶尔，像在乌尔姆，它提供了主导的形式姿态；在其他案例中，例如艾斯利普市的法院、圣何塞市政厅和阿尔普博物馆，圆的形式被主要用作正交对位的形状。在完成于2005年的圣何塞项目中，迈耶在一个开放的广场里放置了独立的穹顶柱体，他是在暗示传统圆形大厅，在广场上，穹顶柱体与曲线的地面层立面连接以形成办公大厦的高、厚两个向度，它本身是迈耶肌理丰富、三维空间组织娴熟的作品之一。

迈耶已经进入用其他方式迎接更积极的视觉韵律和对位关系的阶段。1998年，在佛罗里达那不勒斯的诺依格包尔住宅的蝴蝶屋顶，不仅表达了迈耶其他项目的轻盈，也表达了一种动感；同时屋顶向上的弯折可能会让人回想起勒·柯布西耶的昌迪加尔高级法院，它仍然更加让人回想起一种飞升的感觉。在普拉亚格兰德度假胜地的别墅设计里，迈耶延续了这个想法，但是在意大利贝加莫的意大利水泥实验室，所有这一切更加豪华，在实验室方案里，巨大的水平延伸的屋顶有一个向上倾斜的面，优雅地把围合结构与动感活力融合起来。在这里，迈耶设计的带有纹理的百叶窗墙是水平的且有层次的，这些墙完全地抵消柱子的垂直推力，也是大型不规则延伸屋顶的恰当表达形式。建筑物固定在地面上，同时，它看起来好像可以在景观中穿行。

迈耶的所有建筑，在某种程度上是关于轻盈、透明和白色属性的尝试，白色在迈耶的手中变得不是中立的而是清晰的、锐利的、有效的色彩。迈耶的白色引起你的注意，不论它是否是对绿色的大自然、反光的玻璃，木炭色的花岗岩，抑或是黄棕色的石材（迈耶在罗马和平祭坛博物馆使用了这种材质）的对抗。白色不是安静的颜色，也不是背景色。迈耶建筑拥有的那种柔软的特性来自非凡的优雅和布局的平衡，不是来自立面的白度。人们被一种诱惑力引导，认为迈耶建筑的宁静和它的白度有关，但是事实并非如此：迈耶的建筑是宁静的，而这份宁静几乎与白色无关。纯白色让一切变得激烈，如果作品中没有一种微妙的和谐感（更不用说节点的完全细化），所有都有可能变得过火。

除了他在表达纹理——经常达到如此的丰富度和深度以至于它们的繁复程度可以媲美巴洛克风格——设计语汇的方法上的持续探索，让迈耶全神贯注于他的作品。近期，还有他的另一个追求，即将这种语汇运用到大尺度的高层项目上。纽约的东江项目和在纽瓦克市的苏玛（马凯特街南部）总体规划不是迈耶第一次尝试城市设计——从他的1973年布朗克斯双子公园住宅设计开始，他就有在城市结构下进行物质空间布局的极强的敏感性，它们是目前为止他接手的最大的综合体设计项目。两个项目都大到足以建立自己的文脉，但是不可以依靠如宝石闪光般的典雅在城市景观中证明自己的存在。在不同情况下，高层建筑会创造一个新的城市场所，它的质量将不仅依靠迈耶建筑的轻盈和现存城市环境的沉重之间动人的对比，也依靠组成令人振奋的城市环境的所有元素。

这些都是充满雄心的规划和设计，尺度巨大——东江项目计划用四个塔楼容纳2500个公寓单元——同时，两者都不可能以迈耶原始设计的密度和尺度建设（或者以现在经济环境中的任何形式建设），每一个项目都代表了真正的城市转型的努力。在东江项目中，迈耶从大型开放空间着手开始了他的塔楼（设计），这个大型公共空间西至第一大道，东至河边陆地。两个塔楼的主导轴线垂直于塔楼的边界，也垂直于处于项目北侧的联合国秘书处大厦。水边步行路和几个低层建筑是设计中的都市化元素，设计力图创造新的城市公共空间，并处理好与秘书处大楼形体的关系。

在纽瓦克，离迈耶成长的郊区城镇不远处，他开展了一个更大、更复杂和更能与现存城市建筑融为一体的规划。塔楼的高度和形状更加多样化；没有像秘书处大楼那样的现存元素需要处理，迈耶自由地创造自己的天际线。这些建筑预计会有比迈耶多数小型建筑更加扁平的表皮——他似乎已经发现了纹理表皮细腻的木纹细节在大尺度的建筑中是无法辨认的，这些事情他在麦迪逊广场花园的另外一个双塔设计中并没有意识到，双塔位于曼哈顿市中心，1987年设计，未建成。相比之下，在纽瓦克，若干塔楼的表皮显露斜角，表达结构支撑，这是迈耶新的设计方向。迈耶像密斯一样，拥有一种特殊天赋，他可以在城市结构中布置出美丽、抽象的物体，并且带着令人惊讶的和微妙的能量，让这些物体与城市结构产生联系。在曼哈顿的佩里和查尔斯街公寓塔楼印证了这一点，在特拉维夫市的罗斯柴尔德公寓塔楼也采用了类似的手法。在各种情况下，迈耶用看上去是自我参考的但实际上包含更多内涵的建筑回应复杂和多变的城市肌理。他所有的建筑既是主体对象本身，也是与周边世界的交流客体。迈耶的建筑不仅通过模仿来沟通，还通过微妙的介入来沟通。它们展现了和善、胆魄和精细，这些特性的集合，也是迈耶建筑的神奇之处。

附言
Postscript

弗兰克·斯特拉

除了"年轻人",我不可能把理查德形容为其他人。1958年夏,我在新学校的课上遇到了这个家伙,我看不出他与其他年轻人有什么区别。那是斯蒂芬·格林老师带的绘画班,斯蒂芬老师是纽约杰出的艺术家,早两年间,他一直是我在普林斯顿的老师。斯蒂芬老师邀请我,让我从埃尔德里奇街珠宝店的阁楼出发,在课后和迈耶、斯蒂芬老师的其他学生喝一杯。我记得老师说这些学生相当不错、很活泼、"也不是没有天赋",他加了这几句来美化这个邀请。

我认为我与迈耶如此合得来,是因为斯蒂芬·格林老师把重点集中在我们共同的兴趣点上。由于我和迈耶对斯蒂芬老师的钦佩和对他代表的事业的钦佩,我们可以找到沟通的方法——聊天、找乐、一起工作。

直到今天我们仍然能聊天、找乐、一起工作。在那年夏天结束的时候,我从埃尔德里奇街5号搬到了西百老汇366号。新的空间不比旧的大多少,但是从北面和东面大阁楼窗子进入的光对绘画来讲简直是完美极了。这也是一个让人快乐的空间,变身为精美的艺术聚集地。从传统意义上来讲,自文艺复兴以来,绘画、雕塑和建筑均属于艺术学的范畴。在西百老汇"学术界",我们把摄影加入到传统艺术学的范畴里,所以在瓦特和西百老汇西南角一个20×20的不规则多边形里,就在纯粹午餐餐厅标志上面,我们可以找到负责艺术聚集地的四个人——画家弗兰克·斯特拉、雕塑家卡尔·安德烈、建筑师理查德·迈耶、摄影家霍利斯·弗兰普顿。

我仍能清楚地看到霍利斯为理查德拍的照片,那是一幅在白色卷纸上的照片。理查德穿着三件套的西装,背景全是白色,那显然是成功的年轻建筑师的恰当形象。当霍利斯发送一张我穿着三件套的相似照片给现代艺术博物馆的多萝西·米勒以供她们在"16个美国人"的目录使用时,她喘着气,祈求他再拍一张随意一点的照片。当时我就在想,把一个年轻的画家打扮得像个年轻的建筑师,这不太合适。但是这种年轻人的互通性,这种充满精力、自信地攀登学术高峰的精神让我们受益匪浅。毫不夸张地说,要欣赏理查德作为建筑师的伟大,你必须把他的成就归功于他作为画家和雕塑家的天分。我的意思是,如果没有能力以这样或那样的方式涉足艺术领域的话,我不相信任何伟大或者优秀的艺术作品能够出现。例如,道格拉斯住宅可以被视为一幅画,也就是说,被看作是一幅巨大的连续的抽象表现主义画作,大胆、强烈的白色照亮了场地或者森林植物的墙。这种绘画式的表达源于理查德的天赋和他的工作方式。我从他1959年在我工作室画的红、白、蓝的绘画中看到了这些。我喜欢他的画,虽然我在那时候没有确切地向理查德表达过这个想法,但是我真的对他绘画作品的数量和优秀程度印象深刻。关于那幅画,一直萦绕在我脑海里的是:那些干燥的和明亮的颜色,以及像墙壁一样坚实的建筑品质。在纽约20世纪50年代后期的绘画界,画家在白色干燥区域运用的红色和蓝色宽带状笔触,对我来讲是非常特别的,我们能够很容易地看到绘画中的垂直性,从而预言道格拉斯住宅本身的垂直性。

回想当初,能够找到这些关联,但是当时我们只是就那幅画谈论而已。改变的可能和进步的前景让我们十分开心。在最后,我说"别管它了",迈耶很高兴地同意了。我说"别管它了"是因为我从那幅画得到了一种对完整性的满足感。我当时相信,现在也相信,是完整性赋予艺术作品生命。

感谢理查德所有伟大的工作,感谢那些讨论和学习的美好时光。为理查德,一个真正的艺术家,一个非常本真的纽约文艺复兴时期的人(参与者)而欢呼。

在建筑联盟125次年会之际提交的评论

大学俱乐部

纽约州纽约市

2009年2月9日

理查德·迈耶简介
Richard Meier

理查德·迈耶在康奈尔大学接受建筑设计训练，在1963年创建了自己的工作室。他的项目包括美国、欧洲和亚洲主要民用建筑委托项目，具体包括法院、市政厅、博物馆、企业总部、住宅和私人别墅。他最知名的几个项目就是洛杉矶的盖蒂中心、意大利罗马的千禧教堂、佐治亚州亚特兰大的高级艺术博物馆、德国法兰克福的装饰艺术博物馆、法国巴黎的Canal+电视台总部、西班牙巴塞罗那的当代艺术博物馆。

1984年迈耶被授予普利兹克奖，这个奖是建筑界的最高荣誉。同年，他被选为设计洛杉矶颇为著名的项目——盖蒂中心的建筑师，这个项目1997年12月对外开放，倍受好评。近期，迈耶完成的项目有：德国阿尔普博物馆，罗马和平祭坛博物馆，德国巴登巴登市布尔达收藏博物馆，加州大学洛杉矶分校的布劳德艺术中心，圣何塞市政厅，纽约市佩里街173/176号和查尔斯街165号，艾斯利普市、纽约市和亚利桑那州菲尼克斯市的联邦法院，纽约伊萨卡康奈尔大学的威尔大厅和生命科学技术楼。目前在建的项目有：意大利贝加莫的意大利水泥创新技术中心实验室、以色列特拉维夫市的住宅塔楼、巴黎的圣丹尼斯办公楼和纽瓦克苏玛总体规划。

1997年迈耶获得了AIA(美国建筑师协会)的金奖，美国建筑师协会的最高奖项。同年，日本政府的日本皇室世界文化奖授予迈耶终身艺术成就奖。他是英国皇家建筑师协会和美国建筑师协会会员。他1980年获得了美国建筑师协会纽约分会的荣誉奖章，1998年获得了洛杉矶分会的奖章。迈耶获得了很多奖项，包括30个国家级美国建筑师协会分会的奖项和超过50个地区级美国建筑师协会分会的设计奖项。1989年，迈耶获得了英国皇家建筑师协会的皇家金奖。1992年，法国政府授予他法国艺术及文学司令勋章；1995年他被选为美国艺术与科学研究院院士；2009年，他获得了纽约建筑联盟的总统勋章。他是库珀·休伊特博物馆、罗马的美国学会和美国文学艺术学会的董事会成员，其中美国文学艺术协会于2008年授予他建筑金奖。他获得了那不勒斯大学、新泽西理工学院、新社会研究学院、普瑞特艺术学院、布加勒斯特大学和北卡罗来纳州立大学的荣誉学位。

年代表
Chronology

2004
Giants of Design Award, House Beautiful

Awards for Architecture from the New York Chapter of the American Institute of Architects

National Honor Award for Architecture from the American Institute of Architects

Dedalo Minosse International Prize

Quinquennial Honorary Award for Jubilee Church

Honorary Doctorate of Fine Arts from North Carolina State University

San Diego Federal Courthouse
San Diego, California
Michael Palladino, Partner in Charge

Beach House Miami
Miami, Florida

Southern Florida House
Palm Beach, Florida

Coffee Plaza
Hamburg, Germany

9900 Wilshire
Beverly Hills, California
Michael Palladino, Partner in Charge

Burda Collection Museum
Baden-Baden, Germany
Peter Kruse, Associate Architect

Architecturally, the most prominent part of the expansion, unveiled last month by Michael Palladino of Richard Meier & Partners Architects, is stunningly sophisticated. An impeccably elegant, very slender 22-story tower designed in modernist tradition, the building soars and surprises.
In a major breakthrough that will be much appreciated by the jurors and staff who spend days inside the courtrooms, this tower will be highly transparent. Its glass-curtain wall design, protected on the east and west sides from heat and glare by sunshades, will allow controlled natural light to reach into every courtroom and office. Furthering the architect's design coup in making the courthouse transparent, the tower's base—which is dominated by a jury assembly room designed for up to 700 people—also will be largely transparent. Huge moveable glass panels will open to an elevated, secured terrace, transforming the hall into and indoor-outdoor room probably not seen or experienced by decision-making citizens since ancient Greece.
Ann Jarmusch
San Diego Union-Tribune
22 May 2005

This scheme comprises the exterior enclosure, exterior public spaces, and interior lobby for a twelve-story glass-enclosed condominium located in South Beach. The proposed site was located on 200 feet of beachfront with unobstructed views of the Atlantic Ocean. Floor-to-ceiling glazing and generous terraces characterize the exterior facades, providing panoramic ocean views and copious natural light. The design incorporates a landscaped entrance with reflecting pools and a dramatic waterfall, and a striking lobby composed of a double-height space leading to direct and dramatic views through the building to a private beach club, a pool, and cabanas, all overlooking the ocean.

This outstanding project is a wonderful blending of environmentally sustainable standards and elegant design that will present a graceful gateway to the City of Beverly Hills.
Andy Cohen, FAIA, Co-Chair, 36th Los Angeles Business Journal Los Angeles Architectural Awards

---

The most eye-catching feature of the Sammlung Frieder Burda is a deep declaration of love for the inventor of the rationalist triad composed of the free plan, pilotis, and roof garden. Paradoxical as it might seem, it must be acknowledged that Meier shows incredible originality in finding his own innovative reading of Le Corbusier's design idiom. He constantly succeeds in constructing complex narratives despite using what is basically an obsolete vocabulary almost completely out of use nowadays. For instance, in the case of the new museum the pilotis are radically transformed: no longer just structural features, they are turned into symbolic signs, self-quotation and estrangement to bring the transgressive nature of sign narrative into an inhabitable space. This creates space not just for inhabiting but also for being read and introjected as an abstract experience, an emotion produced by a code capable of translating space into the visionary perception of constantly changing states of mind. This shows how Meier's architecture has an unexpected mannerist vein running through it, working with the legacy of modernity free from its ideological groundings.
Carlo Paganelli
L'Arca
May 2005

Feldmühleplatz Office Buildings
Düsseldorf, Germany
RKW, Associate Architect

Millennium Plaza
Los Angeles, California
Michael Palladino, Partner in Charge

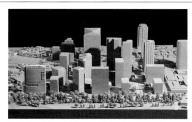 

Mandeville Place
Philadelphia, Pennsylvania
Michael Palladino, Partner in Charge

Project Sandbox
Malibu, California
Michael Palladino, Partner in Charge

 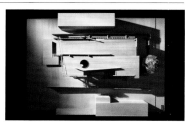

2005
Honor Awards for Interior Architecture,
Architecture, and Housing Design
from the New York Chapter of the
American Institute of Architects

AIA Justice Facilities Review Certificate
of Merit

San Jose City Hall
San Jose, California
Michael Palladino, Partner in Charge

Playa Grande
Dominican Republic

In the spirit of the original plans envisioned by Century City's founders, Millennium Plaza brings an environmentally state-of-the-art residential building to 10000 Santa Monica Boulevard, offering residents the opportunity to rent apartments where they work, shop, and play. The plan calls for 352 apartment units, all with exceptional views, and with 19 units designated affordable housing. The design creates a transition between the multifamily residential neighbors to the east and the commercial high-rises of Century City to the west.

A fifteen-story component of the project faces residential streets, while a thirty-six-story component is compatible with adjacent commercial high-rises. The buildings' materials consist of architectural concrete, aluminum glazing systems, anodized aluminum, and terra-cotta.

The east/west building axis is responsive to urban view corridors from the center city to the river and park. Presenting its slim side to the water, the luxury tower, designed primarily with full-floor units, will preserve public views of the river from Sansom Street. By planning one condominium unit per the majority of floors, the broad north and south facades offer each residence 360-degree views to the city and river. The minimal west exposure limits sun control issues, and the facade ventilators allow natural ventilation at appropriate times of the year.

Palladino is now at work on a thin slab, a vertical mille-feuille of concrete and glass that will hover like a spire over the Schuylkill River Park. Palladino is still tweaking models of the 41-story building, but even in its unfinished state it looks like Philadelphia's best high-rise. You can detect a bit of the PSFS tower in its ancestry, but there is nothing literal in those references—a sign of true creativity.
Inga Saffron
Philadelphia Inquirer

6 February 2006

The design for this small beach house takes advantage of a required view corridor through the property to create a procession from the Pacific Coast Highway to the beach. A raised boardwalk bridges the parking area and the resident's entry, offering framed views to the ocean before entering the residence. Above the two-story entrance, a skylight 10 feet in diameter is designed to open to the sky and act as a natural ventilation chimney for the entire residence. Large areas of south and east-facing glass are shaded with vertical wood louvers that are spaced to manage solar heat gain.

Playa Grande comprises over 2,000 acres of mostly undeveloped land along the northern coast of the Dominican Republic. The master plan for Playa Grande calls for a variety of accommodations and amenities, which includes a total of two hotels, luxury villas, and more modest bungalow-like structures, as well as an equestrian center, retail and cultural space, and the already developed Robert Trent Jones Golf Course. While the Beach Village and Artists' Colony is loosely organized to absorb housing concepts (and other structures) developed by other architects and artists, Richard Meier & Partners has created five prototype plans, each featuring generous studio space: the Butterfly House, the Deck House, the Tower House, the Cube House, and the Modular House. These will range from 50 to 200 square meters, and each will adhere to the use of a modernist vocabulary of materials that will include ferroconcrete and curtainwall glass, as well as apply indigenous materials, particularly wood, in a refined manner that achieves the highest design standards while also responding to the context of the tropical landscape.

Houses in Shenzhen
Shenzhen, China
Sherman Kung, Associate Architect

Italcementi ITC Laboratory
Bergamo, Italy

2006
National Honor Award for Architecture
from the American Institute of
Architects

Ara Pacis Museum
Rome, Italy

UCLA Broad Arts Center
Los Angeles, California
Michael Palladino, Partner in Charge

165 Charles Street
New York, New York

All Saints Church Expansion
Pasadena, California
Michael Palladino, Partner in Charge

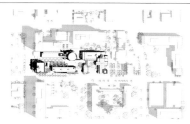

Physically spectacular, Meier's designs for five large houses near the booming Chinese city of Shenzhen are variations on a theme, dotting a hillside with musical precision
and a variety of forms born of the same geometric elements. All the houses have a view of the water, and none of them occupies the top of the hill, which would have been considered bad luck in China. These houses show many of the ways which Meier has continued to explore the same themes throughout his career, and how he has gradually reached a level of maturity
that permits him to take the basic principles even further than he could in his early houses, for example.
Philip Jodidio
Richard Meier & Partners: Complete Works 2003–2008 (Taschen, 2008)

It was no easy task to devise ways of housing and protecting Emperor Augustus's Altar of Peace while also displaying it to the best advantage. The mayor of Rome knew exactly what he was doing when he decided to commission Meier. This is a serene and silent piece, transparent and magnificently lit. Meier has succeeded very well in articulating the elements of this beautiful showcase. He has managed very effectively to capture the bright Roman light. He has built a most beautiful structure.
The intervention of an American in a landmark in the ancient history of Europe reminds me of the writer Henry James and his beautiful story set in Rome, "The Last of the Valerii." Has any European author ever described the Pantheon in such luminous prose? Has any European architect produced a better showcase than Meier's Ara Pacis Museum? With great integrity and skill, Meier has created a work that comfortably engages in dialogue with the historic city while at the same time exploiting to the full the important monument that presides over it. In masterly fashion, Meier has made tangible what Henry James described as "the illusion on golden air" in Rome.
Alberto Campo Baeza
Richard Meier & Partners: Complete Works 2003–2008 (Taschen, 2008)

Meier & Partners transformed the existing structure into the striking, modern building of architectural concrete, teak and white glass that now graces UCLA's North Campus. The firm added outboard structural buttresses to the west end of the tower,
an alternative to interior shear walls that allows for flexible interior space unencumbered by partitions and a loft-like floor plan perfectly suited to studio practices. School of the Arts and Architecture
Dean Christopher Waterman notes, "It'll be fun to watch how the programs grow into the new spaces."

The eight-story Broad Art Center is a portal to the campus from the north, extending the axis right through the building's lobby to the plaza and the Franklin D. Murphy Sculpture Garden. The first public work by sculptor Richard Serra installed in Southern California, a 14-foot-high, steel torqued ellipse titled T.E.U.C.L.A., dominates the plaza and is also visible from several vantage points in the Broad.
Rachel Benioff
UCLA Magazine
11 September 2006

From the exterior, this new tower conjures up an undeniable sense of déjà vu—but with subtle aesthetic improvements over its twin predecessors. While all three buildings share similarities in scale, form, and skins of glass with extruded aluminum painted white, 165 Charles appears sleeker, more taut and self contained. The streamlining here results, in part, from the treatment of balconies: edged in clear glass and inset on a side elevation, in contrast to the translucently enclosed decks projecting from a front corner of each
Perry Street tower. Charles Street's prime facade, forming a smooth, vertical, uninterrupted screen, like a huge sheet of glass, is split only down the middle, where a continuous spine separates each floor's paired riverfront units.
Sarah Amelar
Architectural Record
May 2007

One hundred years ago, All Saints Church was isolated in a landscape of agriculture. The City of Pasadena has grown around the church and, today, it has an important civic presence in the heart of the city. The master plan for the expansion All Saints is designed to meet the program needs of the church and to facilitate the important contribution the church makes to the life of the city. Just as the existing church courtyard supports existing church buildings, the master plan is designed around a new courtyard that organizes and supports three new buildings. The new buildings are scaled and detailed to develop a compatible figure ground and a continuity of materials and texture, creating a unified campus architecture. The architects carefully studied the needs of All Saints Church and have proposed a master plan and design that responds to those needs, and is compatible with the larger urban context. Great care has been taken to create a scale and proportion for the new buildings that maintain the emphasis and focus on the historic church as the iconic anchor of the All Saints campus. The materials and color palette have been selected and designed to be consistent with the palette of the area surrounding City Hall and to be referential to the materials used in the original church buildings. The new buildings are state-of-the-art, environmentally "green" in design.
Ed Bacon, Rector
All Saints Church,
28 May 2008

Scenario Lane
Los Angeles, California
Michael Palladino, Partner in Charge

Joy Apartment
New York, New York

Malibu Beach House
Malibu, California
Michael Palladino, Partner in Charge

CUT and Sidebar by Wolfgang Puck
Beverly Hills, California
Michael Palladino, Partner in Charge

Za'abeel Palace
Dubai, United Arab Emirates

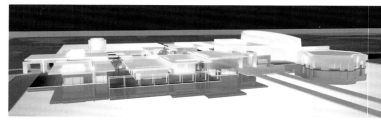

Fukoku Mutual Life Building
Osaka, Japan
Competition

This unique hilltop residential property in west Los Angeles offers a sense of privacy and isolation combined with extraordinary views to the city's reservoir, the Santa Monica Mountains, and framed views to the Los Angeles urban fabric.

The project includes the renovation of an existing residence and construction of a new and separate studio. The large existing house is a terraced and layered architecture that evolved through multiple renovations. Although the building profile and envelope could not be altered, by reconstructing the interior space and adding large areas of glass, every space takes advantage of the extraordinary setting with expansive views and abundant light.

In counterpoint to the ad hoc rambling massing of the existing house, the new studio is a finely crafted, transparent cubic pavilion that cantilevers off the steep property. Architectural concrete site walls anchor the "floating" cube to the hillside, and give structure to the rugged landscape.

This beachfront residence at the north end of the Santa Monica Bay benefits from its south orientation and unobstructed views to the Pacific. A two-story volume of light surrounds the living room, which is at the heart of the plan at the ground level, with the master bedroom suite above. Clear glass, translucent glass, and neoparium (cast glass) panels are selected for their ability to perform in the harsh beachfront environment.

Richard Meier & Partners has given the Beverly Wilshire Hotel, a grand old dame constructed in 1926 and listed on the National Register, a Modernist makeover with "CUT" and "sidebar," a new restaurant and lounge that flank either side of the hotel's French Renaissance-inspired lobby.

CUT, featuring celebrity chef Wolfgang Puck's steak and seafood cuisine, takes advantage of the existing building shell. Arched bay windows inspired the plan's geometry. At the south windows, a painted steel trellis planted with white wisteria extends through the facade and into the dining room, following a semicircular path. A new skylight at the center of the dining room includes translucent suspended glass panels to diffuse and reflect daylight throughout the dining room. The architect removed existing window treatments and, along the west facade, added painted aluminum sunscreens to protect diners from strong afternoon sun. The two-tier dining room is finished with golden-toned quartered white oak floors. Banquette seating of ash and black leather floats on minimal wood supports, while tabletops repeat the floor finish with one-inch oak laminations. Dining chairs with aluminum frames and black mesh seats, designed by Charles Eames, lend classically modern profiles to these spaces.

Across the lobby, the architect gave "sidebar" a more intimate character suitable for a lounge. A bentwood bar, comprised of laminated solid white oak with stainless-steel detailing, provides a dramatic counterpoint within an otherwise rectangular volume. Sofas and settees are upholstered in mohair, combining colors of sage with jewel-toned pillows and black leather ottomans. Vintage table lamps in polished nickel, designed by Josef Hoffman, punctuate these seating groups.
James Murdock
Architectural Record Web site
October 2007

The design for this palace reflects the fundamental programmatic principles of ceremony, tranquility, and political identity, with the overall goal of the design being the contribution of a durable landmark building to the landscape of Dubai. In the context of the palace as an official residence of state, a clear hierarchy of spaces was required, and the plan for the palace emerges from a spatial organization that conveys this hierarchy while creating a dialogue with the rich tradition of building that is to be found in Arabic culture.

The palace is comprised of a series of crystalline geometric volumes floating within a man-made lake. The geometry of both the palace and the surrounding lake is based on a square and the Golden Section, ensuring harmonious proportions for both. A universal module of 9 meters informs every element of the design, from the landscape forms to the design of the floor tiles in the majlis, or assembly rooms. The structure has no beginning and no end; it is designed from the inside out and the outside in so that its form is inextricably bound to the site and its profile expresses a sense of the eternal.

This competition submission for a design for the Fukoku Mutual Life building in Osaka, Japan, envisioned a tower 121 meters in height with twenty-five floors above grade and four floors below grade. The lower floors contained parking and technical spaces and connected to the existing commercial, retail, and transportation network. Situated on a site defined by the high density of the surrounding urban context and its proximity to Osaka Station, the scheme proposed a beacon for the skyline by virtue of its simple geometry and materials.

The architectural intention was to create simultaneously a solid, translucent, and transparent "white" building with minimal maintenance requirements. The plan organization and the main facades defining the southwest corner were designed to be clear and ordered. The technical requirements of solar and heat gain control were integrated into the facade's white "frit" treatment to create a subtle, almost magical effect.

The building was organized vertically around a central, multistory atrium, which featured gallery space as well as undulating benches, glass light wells in the floor, and an illuminated ceiling. The second and third floors were designated for two retail banks, and the fourth and fifth floors were devoted to cultural venues.

Flying Point Residence
Southampton, New York

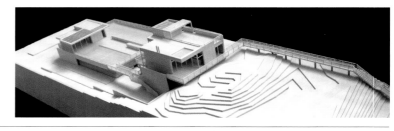

New-York Historical Society
New York, New York

Symphony in the Glen
Los Angeles, California
Michael Palladino, Partner in Charge

2007
European Architecture Prize from
the European Foundation of Culture

Medal of Achievement from the
Philadelphia Art Alliance

Arp Museum
Rolandseck, Germany
Ehrensberger & Oertz, Associate
Architects

Peek & Cloppenberg Mannheim
Mannheim, Germany

Lilium Tower Competition
Warsaw, Poland

This design establishes a new twenty-one-story residential tower above a renovated and expanded home for the New–York Historical Society. The scheme provides a solution for a project that maintains a dual identity: incorporating the expansion of spaces for an august and beloved cultural institution into the creation of a new luxury residential tower with a strong formal profile that is compatible with its rich urban context.

The hybrid nature of the project, with its parallel cultural and residential components, offers the opportunity to develop innovative solutions to the spatial relationship between museum and tower. The tower massing has been made elegantly thin and is set back from the front facade of the museum to create a clear separation from the museum's landmarked architecture. The tower is also located obliquely from the neighboring residential tower on Seventy-seventh Street to preserve that building's views of Central Park and downtown. The new tower structure commences on the sixth floor, just above the proposed museum rooftop addition.

The project delicately separates the public and private spaces while creating special points of intersection, such as a roof garden accessible to residents and museum visitors during special events. The overall design solution provides for the proximity of these separate programs within a single coherent architectural language of intervention.

Located in the Old Zoo Picnic Area of Los Angeles's Griffith Park, Symphony in the Glen presents free classical music presentations in an atmosphere welcoming to children and families. The current setting, originally selected because of the natural amphitheater created by the glade nestled within surrounding hills and trees, informs the architecture.

In 2007 a fire blazed through the park, burning more than 800 acres. The new open-air performance pavilion is designed to engage a landscape that has been graded
to create a welcoming and casual yet acoustically accurate and exceptional outdoor entertainment venue.

Both the gallery spaces with their smoked oak floors and the whole repertoire of Meier's metal staircases and spendidly simple white structural pilotis are gathered together into a fabulously sophisticated white envelope of panels, balconies for viewing and blade-like solar shading. The top light and side light are handled in a similar manner as the Barcelona museum, all extremely careful and well-proportioned. The quality of the construction and finish is simply superb. The sequence of gallery spaces is also exemplary—a kind of modernist ensemble, not a tiring enfilade. Generous seating is provided throughout. The sheer accomplishment this building represents is so inspiring as it reveals just what a great modern architect can deliver when, with a great client, they practice at full stretch, without any inhibitions.
Ivor Richards
Architecture Today
November/December 2007

Conceived with the goal of embedding a sense of civic grandeur as well as enriching the urban fabric of this area of Warsaw, the design for the 121-story Lilium Tower integrates large-scale redevelopment of three sites with concern for the human experience. The reimagining of the Central Station and the school area as well as the existing Marriott site responds to and enriches the quality of life at the pedestrian scale by shifting circulation from below grade to street level, creating new green plaza space for a community experience and erecting two significant civic landmarks in the new railway station and mixed-use tower. The design proposes a combination of hotel apartments and fully residential condominiums as well as two restaurants, conference rooms, and a swimming pool and health club.

A public plaza connects the tower to a public park adjacent to the school area. The plan calls for the demolition of the existing structure and construction of an expanded building accommodating more students and an integrated library, gymnasium, and an assortment of labs and studios.

Also demolished within this plan would be the existing Central Train Station, to be replaced by a 9,200-square-meter world-class train station and a nine-story hotel.

Luxembourg House
Eisenborn, Luxembourg

Rothschild Tower
Tel Aviv, Israel

Bodrum Houses
Bodrum, Turkey

SoMa Newark Master Plan
Newark, New Jersey

Kasa Karma
Beverly Hills, California
Michael Palladino, Partner in Charge

Louise T. Blouin Institute Exhibition
London, England

Meier's latest tower project is a 27-story residential development in the midst of Tel Aviv's "White City," a UNESCO World Heritage site with thousands of Bauhaus-style buildings dating to the Thirties and Forties.

The design incorporates Meier's ideas about light and transparency with advanced technologies to cut water and energy consumption in Israel's heat. Says Meier, "The design of the buildings in the White City made a deep impression on me when I visited Israel many years ago, so to work in this context has been an aspiration of mine for a long time.

"It is my hope that inviting the timeless quality of this city's light and views into an open layout for the residences, a design we haven't seen much here, will bring together the existing elements with a fresh perspective.

"The lightness, transparency and elegance that the design of the Tower intends will integrate well with the lower scaled buildings in the area, and provide a new landmark which complements its Bauhaus predecessors."

Kevin Brass
International Herald Tribune
28 May 2008

This extraordinary residential property is accessed by a 1,000-foot driveway that climbs along a ridge to an elevated pad with views to the entire Los Angeles basin. The architecture and building footprint are designed to celebrate the views and mediate between the topographic geometry of the hillside and the structured garden and landscape of the flat building pad.

The interior planning is designed in a linear organization to allow each living space expansive views to the city, balanced with framed views to scaled private gardens.

The climax of the show, however, comes with three projects not previously shown: the twin residential tower structures at Perry Street, New York, Meier's first built Manhattan work; a 21-house cluster of major holiday villas on the Turkish Bodrum peninsula, which exploit the contours and are predicated on five prototypes, all anchored to a plinth with memorable interior spaces including two-story living areas with views of the seashore. And finally, there is Meier's proposal for The World Trade Center, New York, designed with Stephen Holl, Gwathmey Siegel and Peter Eisenman.

The WTC is one of the least understood of Meier's proposals with its massive order of five 1,000-foot-high crystalline towers interlinked with cantilevered sky lobbies and extensive mix of spatial uses. Withdrawn to the periphery, these towers enclose on two sides a vast landscaped Memorial Square with tree and water ensembles. The whole urban site is homage, and a calm abstract tribute to the contemplative memory the site holds. A selection of Meier's accomplished collages is exhibited in a dense, rich array celebrating his travels, encounters and memorabilia. At the same time there are his sculptures— welded metal abstractions—and his pencil sketches, which accompany the range of exhibits. And, there is his black Knoll International furniture and covetable product designs, especially a pair of pitchers in silver and glass. Finally, there is his splendid grand piano, a magnificent slab-sided instrument, in decadent black lacquer and polished chrome plating, which was in use, with a young lady pianist, for the exhibition's opening party.

Ivor Richards
Architectural Review
March 2008

2008
Twenty-Five Year Award from the
American Institute of Architects

Gold Medal for Architecture from the
American Academy of Arts and Letters

Dedalo Minosse International Prize for
Jesolo Lido Village

Lutz & Patmos Cardigan
designed with Ana Meier

ECM City Tower
Prague, Czech Republic

Pen, Acme Studio

Jesolo Lido Village
Jesolo, Italy

Jack Markuse, High Museum
Watch Reissue

Weill Hall, Cornell University
Ithaca, New York

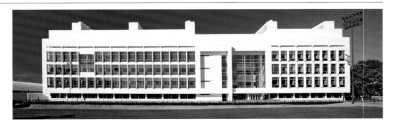

Sharply tailored clothing has often been described by those in the industry as "architectural," but Lutz & Patmos designers Tina Lutz and Marcia Patmos have taken the term to a new level with their latest collaboration. Known for enlisting nondesigners such as Julianne Moore, Liv Tyler and Sofia Coppola to create cashmere pieces for their collection, the New York duo this time turned to starchitect Richard Meier for input. "What we really love about his architecture is that it's similar to our line," Lutz said. "It's clean and not fuzzy or tricky."

With the help of his 27-year-old daughter, Ana, who happens to be a designer, Meier set out to replace the blazer he normally wears with a soft cardigan. The result: a buttonless, V-neck cashmere and cotton cardigan for men, priced at $575, and a women's version with a self-tie belt, for $595. "We wanted very simple lines and a very contemporary feeling," Meier said. "You can wear it as a jacket or as a sweater, or with a shirt or T-shirt. It could be formal or informal."

The cardigans, in stores in January, are available in black, gray and white. T en dollars from each sale will go to Architecture for Humanity, a nonprofit organization through which architects donate their services to needy communities. Meier particularly enjoyed the speed with which his sartorial vision was turned into reality. "I just had a building open in Germany that took 20 years," he said. "This was 20 days. It's a lot simpler."
Marc Karimzadeh
W Magazine
January 2008

Twenty-five years later, and tenants can finally move into the tallest building in the Czech Republic. Prague's City Tower, designed by Richard Meier & Partners, will officially open in March after sitting unfinished for decades. The tower is part of ECM city, an ongoing urban renewal project with a master plan also spearheaded by
Richard Meier & Partners.

The 530,000-square-foot-glass structure, located in Pankrác, part of the Prague 4 District, is 358 feet tall. It was originally planned for Czechoslovak Radio, but in 1983 the steel skeleton was abandoned. ECM Real Estate Investments purchased the property in 2000 and construction began once more in 2004.

City Tower features 27 above ground and three underground floors. Design highlights include floor-to-ceiling glass offering views of the historic city and a VIP restaurant and conference facility, located on the 27th floor. The project has met with some public controversy due to its size—eight other buildings are in development.
Mairi Beautyman
Interior Design
January 2008

It can take five or fifteen years to complete a museum, so designing small scale objects can be very gratifying. When my friend and collaborator Massimo Vignelli told me he was designing a pen for ACME Writing Tools, I was intrigued. I knew the founders of ACME, Adrian Olabuenaga and Lesley Bailey, and I realized what a challenge it would be to design an instrument that is an indispensable element of my daily life. This pen looks good. It feels good. There's no fuss or fancy as with many luxury pens, and it's white. The fine point and inkflow are such that I can use it to draw on napkins or tracing paper, write a letter or sign one of my collages. They only made 500. I'm glad I got one.
Richard Meier

Blackbook Magazine
April 2008

The project is at the northeastern end of Jesolo on the site of the former colony Monte Berioc, which was erected in 1949 and until a few years ago was visited by hundreds of young people from the surrounding area. The fresh and cheerful atmosphere that one notices on entering the site is truly pleasant. Finally it has proved possible to introduce a little light and air into the densely cemented development structure. The colour white—which in addition to Meier's clear formal language was one criterion for the choice of architect—is also a successful atmospheric element. The core of the housing development is formed by a generously dimensioned atrium with swimming pool and sunbathing areas that is closed on three sides by three-storey paired houses. On the internal side the houses are connected by a shared facade element that provides space for the balcony terraces. The fourth side, facing the main street, is left open and is protected by an earth embankment that will be planted with trees. The generous amount of glazing towards the courtyard and street side brings sufficient light into the apartments, which have highly efficient floor plans. A residential module consists of two bedrooms, a living/dining room, sanitary facilities as well as balconies with external staircases. Those living on the ground floor have a small garden. Translucent curtains have been fitted in all the apartments. They filter the light in a most pleasant way and establish the necessary distance to the unspectacular surroundings. The paired construction system offers an alternation between volumes and empty spaces. There is an underground garage beneath the entire complex. All in all the new housing complex suggests a hint of luxury, which will certainly exert a long-term effect on the surroundings. It is most desirable that more projects of this quality be erected in the future—the Adriatic region needs them badly.
Roland Gruber
Arkitektur Aktuell
March 2008

This reissue of the watch designed by Richard Meier for Markuse commemorating the High Museum represents the perfect balance of space and lighting. Meier's interpretation is reflected in his use of a partially opaque dial with a "light up" button for night viewing which balances against the Museum's architecture of space and lighting. It is made of brushed stainless steel and comes with either a stainless steel mesh band or a soft white leather band.

Richard Meier & Partners' green design for the $162 million facility includes a living roof covering over 50 percent of the building. The roof absorbs rain water and provides insulation; light temperature and air flow are regulated via motion detectors.
The center also boasts high-tech mechanical systems projected to save energy at a rate of more than 40 percent above American Society of Heating, Refrigerating and Air-Conditioning Engineers (ASHRAE) standards, as well as systems to minimize light pollution, water use, and material emissions.

"Environmental sustainability and energy efficiency have been fundamental to both the design approach and construction process of Weill Hall," says Renny Logan, Associate Partner in charge of the project. "LEED Gold Certification recognizes Richard Meier & Partners' and Cornell's substantial commitment to the creation of environ-mentally responsible buildings that make the most efficient use of the earth's resources."
Nicholas Tamarin

Interior Design
22 October 2008

Watch, Pierre Junod

Rickmers House
Hamburg, Germany

Pacific Heights Residence
San Francisco, California
Michael Palladino, Partner in Charge

Malibu Residence
Malibu, California
Michael Palladino, Partner in Charge

Gagosian Gallery Addition
Beverly Hills, California
Michael Palladino, Partner in Charge

Handsmooth House
Ipsden, Oxfordshire, England

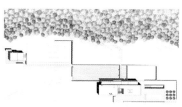 

Swiss watch brand Pierre Junod has launched a new design by architect Richard Meier. The watch has a sapphire crystal, 316 L stainless steel case and an ETA Swiss quartz movement. The 12 squares on the face and the hands are visible in the dark.
Rose Etherington
Dezeen Design Magazine
23 January 2008

This terrace house overlooks an ornamental lake in the center of Hamburg. The 9-meter-wide building is divided into four levels above grade, with living space on the ground floor, guest room and study on the second floor, children's rooms on the third, and a set-back penthouse, containing the principal bedroom with bathroom and roof terrace, above. Virtually the entire site has been excavated to accommodate staff quarters, storage, and a garage.

The building has a classic terrace house plan with a generous, elevated entrance to one side of the forecourt on axis with a top-lit stair/elevator hall within. This system of access runs through the full height of the building. Wherever possible, bathrooms are housed within the width of the service core carrying the vertical circulation.

The house's three free facades are screened by two-story translucent white glass panels with strategic apertures providing views and direct daylight. The load-bearing party wall on the west is structurally tied at each level to the freestanding concrete columns running just inside the eastern face of the building. A paved rear garden opens onto a glass-bottomed water basin that illuminates the garage below.

Located on the seventh floor of an existing 1950s-era San Francisco residential mid-rise, this newly renovated and redefined Pacific Heights apartment responds to the ever-changing daylight qualities and views of the San Francisco Bay, a significant personal art collection, and the clients' desire for the elegance and warmth of a well-crafted and comfortable home.

By reengineering an existing shear wall, two sizable openings were achieved, completely transforming the main circulation spaces that surround the building core.
These galleries connect on axis with the daylight and views of the main living and dining areas, redefining the spatial order of the apartment. The more private study and bedroom spaces are also accessed from this gallery.

At the heart of the plan, the building core is sheathed in continuous horizontal wood panels, creating a central cabinet that one enters through and circulates around. This and all other apartment cabinetry is fashioned from matched panels of light-figured Anigre veneer. Wide-plank, quartered, white oak floors are installed over a radiant heat system.

This beachfront residence at the north end of the Santa Monica Bay benefits from its south orientation and unobstructed views
to the Pacific. A two-story volume of light surrounds the living room, which is at the heart of the plan at the ground level, with the master bedroom suite above. Clear glass, translucent glass, and neoparium (cast glass) panels are selected for their ability to perform in the harsh beachfront environment.

In 1995 the Gagosian Gallery in Beverly Hills opened to critical acclaim and has become a landmark for the presentation of contemporary artwork.

The gallery expansion will create additional exhibit space for large-scale works of art with a quality of light and space that is consistent with the existing gallery. The new gallery integrates existing wood bowstring roof trusses with skylights along the north and south walls to fill the space with a lively and changing daylight.

The street frontage will double in length and extend the palette of smooth white plaster and clear and translucent glass. The facade is designed to slide open to allow the gallery to be exposed to the street and offer an encounter with the art to the
Beverly Hills community.

The roofscape of the new gallery will be animated with skylights and a roof terrace
to be used for receptions associated with exhibit openings.

Fifth Avenue Apartment
New York, New York
with Rose Tarlow Melrose House

Forstmann Little and Co.
and IMG Worldwide Headquarters
New York, New York

2009
President's Medal from the
Architectural League of New York

On Prospect Park
Brooklyn, New York

Saint-Denis Office Development
Saint-Denis, France
Mas & Roux, Associate Architects

Transbay Block 8 Redevelopment
Competition
Los Angeles, California
Michael Palladino, Partner in Charge

Beethoven Festspielhaus Competition
Bonn, Germany

Surrounded by a bevy of New York's traditional brownstones and architectural monuments Richard Meier's latest glass condominium building is certainly an eye-catching addition to the prestigious location from which the building takes its name, On Prospect Park.

Situated perpendicular to the park's spectacular arched main entrance the 15 storey building benefits from its location both by standing out in landmark fashion and with spectacular views achieved from within. Meier's design makes the most of these views, which give either a park or a cityscape view, by utilising floor-to-ceiling and wall-to-wall glazing. Acoustic and solar performance is ensured by the use of advanced low-E insulated lamination on the glass and an overall bright and airy design is created.

It is not merely the shell of the building that Meier has designed but features including the kitchen millwork in the 102 condominiums. Large balconies with ipe hardwood decking, terraces with jet mist granite and solid oak flooring throughout ensure that the building teams with the kind of quality indulgence that you would expect from a Pritzker prize-winning architect.
Niki May Young
WorldArchitectureNews.com
23 July 2008

The massing and architectural concept for Block 8 is designed to present an original and timeless addition to the San Francisco skyline combined with a scale and cadence at the street level that supports pedestrian traffic and urban living. The four street elevations will share a common scale but will be articulated to express the specific use and programming at each facade.

The residential tower is conceptualized as a layering and shearing of vertical transparent and opaque planes. The tower is firmly anchored against the mid-rise plinth and cantilevers vertically utilizing the elevator shafts to buttress the tallest floors of the tower. The vertical proportion of the architecture is enhanced by a 40-foot vertical extension of the southwest curtainwall with integral photovoltaic cells. All residences in the tower will benefit from the thin proportions of the architecture with daylighting and natural ventilation offered to every room.
At the heart of the Block 8 Development is a 5,400-square-foot landscaped open space. Childcare support for the community is developed around this ground-level garden, which extends to the roofscape of the plinth architecture and will be made accessible to all residents of Block 8.

Richard Meier builds parallel to the Rhine tightly interconnected new structures. They are situated on the East side of a new design for a city park, the "Beethoven Plateau."

He carries out a clear concept; at its basic principle being orthogonal it uses the entire width of the site. In its depth, meaning toward the city, it allows for much space dedicated as park landscape. The location of the large auditorium is clearly recognizable. The roof is lifted to accommodate sufficient volume for excellent acoustics and the white metal of the exterior cladding contrasts the expansive glass surfaces of the other parts of the building. The lower level is set back, creating the effect of a box sitting on top. During the day metal reflects the daylight, in the evening the building glows from within. The small auditorium is
located toward the park.
General-Anzeiger
31 January 2009

理查德·迈耶及其合伙人事务所模型博物馆
Richard Meier & Partners Model Museum

理查德·迈耶及其合伙人事务所模型博物馆
Richard Meier & Partners Model Museum

纽约州长岛市，2007

这个原始展览空间占地334.5平方米，展览着20世纪60年代至今的作品——包括在康涅狄格的史密斯住宅的设计初稿模型（1965），这是确立迈耶声望的早期作品之一。工作室最醒目的模型是盖蒂中心（1984—1997）大尺度的表现模型和研究模型，盖蒂中心被广泛地认为是最能体现迈耶实力的项目。博物馆展出的其他杰出作品的模型还包括霍夫曼住宅（1966）和荷兰皇家造纸厂总部（1987）等。学生和设计爱好者感兴趣的还有未建成项目的模型，例如法国雷诺总部设计（1981）、美国康奈尔大学本科生宿舍（1974）、20世纪福克斯餐厅和电视新闻工作室（1990）。展厅中的长椅由迈耶设计，这是斯托·戴维斯办公家具系列的部分作品。博物馆也展出了迈耶的墙壁和地板系列雕塑作品。这些雕塑由盖蒂中心模型所使用木材的边角料制成，这些木材用不锈钢镶铸，重新组装。它们以德国南部的巴洛克教堂的名字命名。

参观者在2007年5月被非正式地邀请到这里参观，在博物馆开放的最初四个月里，人们对展廊投入了巨大的热情以至于这个项目已经演变成一个季节性的展览。因为气候对模型的影响，仓库在冬天不对外开放。

参考文献
Bibliography

General
Jodidio, Philip. Meier. Taschen, 2009.
Aguilar, Andrea. "Las casas blancas." El Pais Semanal, 10 May 2009. Silva, Horacio. "Richard Meier." Pin-Up Magazine, Spring/Summer 2009, pp. 18–25.
Conforti, Claudia. Richard Meier. Federico Motta Editore, 2009.
Huxtable, Ada Louise. "The Meier Superstyle." On Architecture: Collected Reflections on a Century of Change. Walker & Company, 2008, pp. 60–67.
Jodidio, Philip. Meier: Richard Meier & Partners, Complete Works 1963–2008. Taschen, 2008.
"My Desk: Richard Meier." Vanity Fair, December 2008, p. 166.
Farkas, Alessandra. "Interni: Libro E Architetto." Corriere della Sera, September 2008, pp. 166–173.
Meier, Richard. "Less Is More." Esquire, September 2008, 152–153.
Shields, Jody. "Richard Meier." L'Uomo Vogue, April 2008.
Richard Meier Houses and Apartments. Rizzoli, 2007.
Richard Meier Museums. Rizzoli, 2006.
Goodman, Wendy. "Prince of the City." House & Garden, November 2007, pp. 100–107.
"Architects' Homes." a+u, March 2007, pp. 86–89.
Gardner, James. "Elegance Meets Efficiency." The New York Sun, 26 September 2005, pp. 12.
Plumb, Barbara. "Rifugio D'Architetto." Casamica, November 2005, pp.22–27.
Boodro, Michael. "Richard Meier 12 things He Can't Live Without." Elle Décor, December/January 2005, pp. 84.

165 Charles Street
Green, Penelope. "The Stylish Terrarium: Decorating Under Glass." New York Times, 11 October 2007, pp. F1, 6.
"165 Charles Street." Casa Brutus, October 2007, p. 71.
Amelar, Sarah. "165 Charles Street: New York City." Architectural Record, May 2007, pp. 218–221.
Barbanel, Josh. "The Sizzling Luxury Market." New York Times, 25 February 2007.
Richard, Pascale. "Manhattan: Luxe a tous les etages." Le Monde 2, 14 October 2006, pp. 54–57.
Jana, Reena. "Richard Meier: Maximizing Minimalism." Business Week, 11 April 2006
Yoshida, Mika, and David G. Imber. "Living in Art: 165 Charles Street." Casa Brutus, February 2006, pp. 176.
Maggi, Laura. "Case D'Autore: Quando gli architetti disegnano residenze al top." Elle Décor, January–February 2006, pp. 103–106.
Anderson, Kurt. "The Five Best New Buildings." New York Magazine, 19 December 2005, pp. 72
De Almeida, Eduardo. "Torre envidra da para visualizar Manhattan." Projeto Design, 5 December 2005, pp. 74–77.
Van Deusen, Amy. "Tastemakers: Architecture." Forbes, 15 November 2005.
Robledo, S. Jhoanna. "The Height of Fashion." New York Magazine, 7 November 2005, pp. 42–50.
Bernstein, Fred A. "Come Right In." Interior Design, September 2005, pp. 256–267.
Treffinger, Stephen. "Room to Improve." New York Times, 4 August 2005, p. F2.
Bartl, Alexander. "Aus Gutem Hause." Elle (German Edition), August 2005, pp. 78–80.
Goldberger, Paul. "A New Beginning." The New Yorker, 6 June 2005.
Neuman, William. "New Home Off Broadway." New York Times, 29 May 2005, section 11, p. 2.
Keil, Branden. "Gimme Shelter." The New York Post, 28 May 2005, p. 8R.
Gardner, James. "How much for a World-Historic Home?" The New York Sun, 2 May 2005, p. 16.
Robledo, S. Jhoanna. "A Million-Dollar Room." New York Magazine, 11 April 2005, p. 68.
Pogrebin, Robin. "For Act II, Architect Gets More Hands-On." New York Times, 11 April 2005.
"One More Twin." Shtab-kvartira, April 2005, p.134.
Lebon, Manuel. "165 Charles Street Condominios de Autor." Complot Magazine, March 2005.
Le Sommier, Regis. "L'immobilier toujours plus haut." Paris Match, March 2005.
Taylor, Chris. "Get Smart." New York Post Home, 8 January 2005, pp. 8R–9R.
Miranda, Agustina Paez. "Arquitectura de Autor." Estilos de Vida, December 2004, pp. 28–31.
Schoeneman, Dehorah. "Condo Couture." New York Magazine, 1 November 2004, pp. 30–35.
Cutler, Steve. "Richard Meier." New York Living Magazine, November 2004 pp. 14–17.
"The Price of Living." Metro, 14 September 14 2004, p. 13.
Neuman, William. "Studios Breaking 7-Figure Barrier." New York Times, 12 September 2004, Section 11.
Ouroussouff, Nicolai. "The New York Skyline." New York Times, 5 September 2004, section 2, page 1, column 2.

Fletcher, Mansel. "The Best ... Penthouse in the World." Esquire, September 2004.
"Richard Meier Charles Street Apartments." a+u, August 2004, pp. 22–25.
"Downtown Living." New York Living, August/September 2004, pp. 32–37.
Neuman, William. "A Posse of the Fabulous Braves the Far, Far West." New York Times, 11 June 2004.
Forsythe, Jason. "Nestling in Lofty Real Estate Clouds." International Herald Tribune, 25 June 2004.
Freedman, Lisa. "The Artist and the Architecture ... or the Architect and the Art." Financial Times, 5–6 June 2004, p. W14.
Belogolovsky, Vladimir. "White Architecture of Richard Meier." Sovremenny Dom, 26 May 2004, pp. 26–33.
Holztman, Anna. "Lifestyles of the Rich (Buyers) and Famous (Architects)." Architects Newspaper, 2 November 2004, pp. 8–10.
"Meier Way." Wallpaper, May 2004.
"Hadid's New Work ... In Meier's Neighborhood." New York Magazine, 19 April 2004.
Sherman, Gabriel. "My, Oh Meier!" New York Observer, 13 April 2004.
Del Ponte, Ivan. "Architecture: A New York, un inno alla limited edition:I tre "condo" de Richard Meier." Casa Vogue, April 2004, p. 28.
"Richard Meier Apartments for Sale as Limited Editions." The Art Newspaper, April 2004.
"Turm-Trio: Richard Meier baut Luxusappartements in Manhattan." Baunetz, 24 March 2004.
Sherman, Gabriel. "Three's a Crowd." The Financial Observer, 22 March 2004, p. 13.
Anderson, Lincoln. "Selling Architecture as Art on the Waterfront." The Villager, 17–23 March 2004.
Rich, Motoko. "The Art of Selling Luxury Condos as Art." New York Times, 11 March 2004, p. F8.

9900 Wilshire
"Next LA/Merit." Form, October 2007, p. 43.
Smith, Clif. "Doing Something Really, Really Good Sometimes Pays Really, Really Well." Beverly Hills Courier, 13 April 2007, p. 39.
Vincent, Roger. "In Beverly Hills, High End Indeed." Los Angeles Times, 11 April 2007.
Seitz, John. "Beverly Hills Property Sold For Half Billion Dollars in U.S. Record Sale." Beverly Hills Courier, 10 April 2007.
Kotler, Steven. "Eco Buildings: Warners, WMA Set a Green Example." Variety, 30 November 2006.
Zenarosa, Michelle. "Architect Unveils Proposal For Robinsons-May Site." Beverly Hills Weekly, 8 December 2005.
Vincent, Roger. "Condo Developers Make Environmental Pitch." Los Angeles Times, 7 December 2005, pp. C1, C9.

Ara Pacis Museum
Casati, Cesare Maria. "It Sometimes Happens." L'Arca, October 2008, p. 1.
Ouroussoff, Nicolai. "An Oracle of Modernism in Ancient Rome." New York Times, 25 September 2006, pp. E1, E4.
Bernstein, Fred. "The Rise of His Roman Empire." Culture & Travel, October/November 2007, pp. 77–81.
Rosa, Giancarlo. "Ara Pacis: The Finished Work." Frames Architettura Dei Serramenti, September 2007, pp. 62–67.
Tita Puopolo, Sonia. "High Fashion:The Victorious Valentino." Haute Living, September 2007, pp. 88.
Gerfen, Katie. "Ara Pacis Opens to Controversy." Architecture, May 2006, p. 28.
D'Alessio, Diamante. "La lunga guerra dell'Ara Pacis." Panorama, 5 August 2004, pp. 66–70.
Nolan, Linda Ann. "Emulating Augustus." Archeology Odyssey, May/June 2005, pp. 38–47.
Colonnelli, Lauretta. "Ara Pacis, 'progetto Meier da annullare.'" La Corriere della Sera, 5 November 2003.
Colonelli, Lauretta. "Attorno all'Ara Pacis strutture ..." La Corriere della Serra, 6 March 2003, p. 50.
Ruiz, Cristina. "Peace Breaks Out over the Ara Pacis." The Art Newspaper, March 2003, p. 22.
Ouroussoff, Nicolai. "Rome Rises Again." Los Angeles Times, 9 February 2003, p. E1.
Giuliani, Francesca. "Ara Pacis: The Second Phase has Started." La Repubblica, 29 January 2003.
La Rocca, Eugenio, and Orietta Rossini. "L'Ara Pacis ha vinto la Guerra." Il Giornale dell'Architettura, January 2003, pp. 1, 22.
Cinanni, Maria Theresa. "Ara Pacis, riprendono ..." Il Nuovo, May 2002.
Colonelli, Lauretta. "Ara Pacis, le modifiche di Meier." La Corriere della Serra, 11 April 2002.
Pearson, Albert, and Catherine Siphron. "A Modern Classic." SOMA, December 2001/January 2002, p. 42.
Quintavalle, Arturo. "Ara Pacis, piu leggerezza per il progetto Meier." La Corriere della Serra, 5 December 2001.
Casciani, Stefano. "Richard Meier's Ara Pacis Museum." Domus, November 2001, pp. 144–145.

Wise, Michael. "Feud Over Fascist Legacy." ARTnews, April 2001, p. 76.
Wise, Michael Z. "Dictator by Design." Travel + Leisure, March 2001, pp. 102–108.
Stanley, Alessandra. "Colorful Characters Lurk Around Monuments." New York Times, 29 June 2001, p. A4.
Ruiz, Cristina. "Richard Meier to Obliterate Mussolini's Work." The Art Newspaper, May 1999, p. 6.
Fitzgerald, Ian. "Augustus' Resting Place." History Today, May 1998, pp. 29–30.
Giuliani, Francesca. "I Took away the Wall and Put a Fountain." La Repubblica, 18 April 1998.
Muratore, Giorgio. "Roma non è l'America: troppo disimpegno storico nel progetto di Meier per l'Ara Pacis." Il Messaggero, 10 March 1998.
Madeo, Liliana. "Provinciale, devastante, troppo costoso: il critico d'arte boccia il progetto di Meier a Roma." La Stampa, 8 March 1998, p. 23.
"Sistemazione museale dell'Ara Pacis, Roma." Zodiac 17, May 1997, pp. 128–133.
Casalini, Simona. "Rinascimento in chiave romana." La Repubblica, 21 June 1996, p. 5.
Pullara, Giuseppe. "L'Ara Pacis 'made in Usa.'" Corierre della Sera, 20 June 1996, p. 49.
Maestosi, Danilo. "Un museo di vetro per l'Ara pacis." Il Messaggero, 20 June 1996.
"Meier scopre l'Ara Pacis del 2000 un nuovo museo da 21 miliardi." La Repubblica, 20 June 1996.
Diehl, Ute. "Shadow Plays and Architectural Light." Frankfurter Allgemeine Zeitung, 13 June 1996.
Guerzoni, Monica. "Meier va all'Ara Pacis." Corriere della Sera, 25 April 1995, p. 32.

Arp Museum
Richards, Ivor. "No Compromise: Richard Meier's Arp Museum Completed after Three Decades." Architecture Today, November/December 2007.
"Großer Bahnof." Frankfurter Allgemeine Zeitung, 18 September 2003, p. 36.
Rasch, Jaqueine. "Schneeweiße Kiste fur die Kunst: Richard Meier prasentierte seinen endgultigen Entwurf des Arp-Museums." An Rhein und Sieg, 17 September 2003, p. RKU01A/1.
"Startschuss fur das Arp-Museum in Remagen." General-Anzeiger, 17 September 2003, p. 1.
Meyer, Werner. "Eine neue Perle fur die rheinische Museenkette." General Anzeiger, 17 September 2003, p. 5.

Powell. Nicholas. "The Battle over Arp." Art News, June 2001, pp. 106–113.
Vellender, Frank. "A Clear Confession for the Architecture of Richard Meier." Bonner General Anzeiger, 28 April 2000.
"Fight Over the Works of Hans Arp." Minzer Rheinzeitung, 14 August 1997.
"A License for Prinitng Money?" Allgemeine Zeitung, August 1997
"Arp-Museum 1999 eröffnet." Bonner Generalanzeiger, 2 April 1997.
"Hans Arp-Museum in Rolandseck." Kunstzeitung, February 1997.
Maurer, Caro. "Beautiful View: Richard Meier and the Arp Museum." Bonner Generalanzeiger, 12/13 October 1996, p. A13.
"Arp Museum, Rolandseck." Bauwelt, March 1996, p. 426.
Sainz, Jorge. "El castillo blanco." CCA & V, February 1993, pp. 108–112.
"Mit Weitsicht." Md—International Magazine of Design, February 2008, pp. 34–39.
"The New Arp Museum Comes into Being ..." August 1997, p. 112.
"Contemporary Rhine Castle." Abstract, February 2008, pp. 24–29.
Futagawa, Yukio, ed. "Arp Museum." GA Document 100, 2008, pp. 64–77.

Burda Collection Museum
Knöbl, Sandra. "Mit Licht für—mit—neben Kunst." Architektur. February 2006, pp. 44–51.
Yoshida, Nobuyuki. "Frieder Burda Collection." A+U, November 2005, pp. 83–93.
"Richard Meier & Partners Architects Burda Collection Museum." Casabella, July 2005, pp.14–23.
Galloway, David. "Building for Art." Art in America, June/July 2005, pp.140–145.
Paganelli, Carlo. "Un pensiero in progresso" L'Arca, May 2005, pp. 2–7.
Forgey, Benjamin. "La perfeccion, en pequena escala." Arq, 19 April 2005, pp. 12–15.
Bennett, Paul. "A Delicate Imbalance." Architecture, March 2005, pp. 46–51.
Forgey, Benjamin. "Windows of Opportunity at a German Museum." Washington Post, 13 February 2005.
Bernstein, Richard. "A Personal Vision, with a Fortune to Match, Creates a New German Museum." New York Times, 20 Jan 2005, p. E3. Reprinted as "Expressionism's New Home." International Herald Tribune, 22/23 January 2005.
"Das neue Sammlermuseum von Freider Burda in Baden-Baden." Museum Aktuell, September 2004.
"Die Baden-Badener Brucke." Art Info, November–December 2004.

Global Update." Architecture+, December 2004, pp. 9–17.
Jacob, Werner. "White House." Deutsches Architektenblatt, December 2004.
Restorff, Jorge. "Streifzug durchs Kunstjahr." Kunstzeitung, December 2004.
Feeser, Sigrid. "Ein Hartetest fur die Malerei." Staatsanzeiger fur Baden-Wurttemberg, 2 November 2004.
"Sammlung Frieder Burda in Baden-Baden Richard Meier & Partners." Baumeister, November 2004.
"Next Artbeat—The New Season." Travel + Leisure, November 2004, pp. 173–178.
Meier-Grolman, Burkhard. "Seine Kunst braucht keine Steckdose." Sudwest Presse, 30 October 2004.
Krause, Christiane. "Burda-Museum erlet Riesen-Andrang." Badische Neueste Nachrichten, 25 October 2004.
"Geduldsprobe vor Museumsbesuch." Badisches Tagblatt, 25 October 2004.
Patzer, Von George. "Kunst ohne Stecker." Die Tageszeitung, 25 October 2004.
Hoffmann, Gabriele. "Ein Bruckenschlag in Badeb-Baden." Neue Zurcher Zeitung, 24 October 2004.
"Modernes Schmuckkastchen im alten Park." NZZ am Sonntag, 24 October 2004.
Von Hammelechle, Sebastian. "Liebenswert private Momente." Welt am Sonntag, 24 Octcober 2004.
Wagner, Thomas. "Gute Beispiele verderben schlechte Sitten." Frankfurter Allgemeine Zeitung, 23 October 2004, p. 31.
Burhr, Von Elke. "In der Villa." Frankfurter Rundschau, 23 October 2004.
Kolgen, Birgit. "Auf der Suche nach der Zetlosigkeit." Schwabische Zeitung Leutkirch, 23 October 2004.
"Richard Meier: Museum mit zeiltloser Wirkung." Badisches Tagblatt, 21 October 2004.
Glanz, Alexandra. "Ich bezahl alles." Koppelstatter, 22 October 2004.
De Righi, Roberta. "Ein Geschenk aus Licht und Farben." Munchner Abendzeitung, 22 October 2004.
Von Neubeck, Hanskarl. "Optisches Reizklima." Sudwest Presse, 22 October 2004.
"Sammlung Frieder Burda: Baden-Baden Feiert Die Eroffnung im Newbau von Richard Meier und in der Stastlichen Kunsthalle." Badisches Tagblatt, 21 October 2004.
Von Neubeck, Hanskarl. "Oprisches Reizklima." Hannoversche Allgemeine Zeitung, 21 October 2004.
Von Kupke, Susanne. "Einmalig in Europa: Mazen baut Museum." Aachener Nachrichten, 20 October 2004.
Von Kupke, Susanne. "Burdas Kunstgeschenk an die Kurstadt." Frankfurter Neue Presse, 25 October 2004.
Koldehoff, Stefan. "Lerne, neu zu lernen." Suddeutsche Zeitung, 13 October 2004.
"Magnet Burda-Museum." Der Sonntag, October 2004.
Von Tilmann, Christina. "In Baden-Baden sind die Brucken hell." Der Tagesspiegel, 23 October 2004.
Sammlung Frieder Burda. Der Bau von Richard Meier. Hatje Cantz Verlag, 2004.
Toldsorf, Stefan. "Ein Meier fur Burda." Die Welt, 4 December 2003.
"Masstoleranz wie in Goldschmiede." Badisches Tagblatt, 11 October 2003.
Langer, Karsten. ""Burda im Doppelpack." Manager, 8 August 2003.
Marino, Silvina. "Exaltacion de Blanco." Clarin, 14 July 2003, pp. 7–8.
Newhouse, Victoria. "Art Matters." Architectural Digest, March 2003, pp. 60–69.
"Bilder Einer Leidenschaft: Die Sammlung Frieder Burda in Baden-Baden." Festspielhaus, February 2002, pp. 32–35.
Hartmann, Serge. "Baden s'offre Burda qui s'offre Meier." Dernieres Nouvelles D'Alsace, 16 October 2001, p. 6.

East River Master Plan
Bagli, Charles. "Towering Vision by Developer Stirs East Side." New York Times, pp. B1, B4, 15 November 2007.
Keil, Brad. "'Beast' Side: Architect's River High-Rise Plan." New York Post, pp. 8. 11 November 2005.

ECM City Tower
"Prague Peak." 100% Office, March 2005, p. 27.

Eli & Edythe Broad Art Center UCLA
Waring, Lisa. "Building on Their Passion." UCLA Great Futures, Fall 2000.
Frank, Peter. "For Art's Sake." UCLA Magazine, Winter 2000, p. 45.
Brinkmann, U. "Fakultat der Kunste und Architektur des UCLA." Bauwelt, 6 October 2000, p. 2.
Anderton, Frances. "The Suavely Familiar vs. the Daring of the Internet Age." New York Times, 21 September 2000, p. F3.
Ouroussoff, Nicolai. "The Mark of Meier Will Make UCLA Arts Center Very Visual." Los Angeles Times, September 8, 2000, pp. F1, F22.

Feldmühleplatz Office Building
Merkel, Jayne. "The Fluid Office." Architectural Record Review, November 2003, pp. 7–14.

Fifth Avenue Apartment
Giovannini, Joseph. "New York Attitude." Architectural Digest, March 2009, pp. 84–91.

Jesolo Lido Village, Condominium, and Hotel
"Richard Meier & Partners, Jesolo Lido Village." L'Arca, April 2008, pp. 24–29.
Gruber, Roland. "Anlage Jesolo Lido Village, Italien." Architektur Aktuell, March 2008, pp. 114–125.
Molteni, Enrico. "Lunghe Linee Americane." Casabella, March 2008, pp. 74–81.
Di Alberti, Luigi. "Nebbia sulle torri di Meier." Il Giornale Dell' Architettura, January 2006.
Cagnassi, Giovanni. "A rischio le tre torri di piazza Milano." La Nuova, 13 May 2005.
"Ancora Richard Meier." Abitare, April 2005.
"Plans Announced for Jesolo Lido Complex." Projection, January 2005, p. 4.
Cibin, Fabrizio. "L'architetto Meier incontra il Sovrintendente e gli ilustra il suo 'Jesolo project.'" Il Gazzettino, 13 May 2004, p. 16.
Carafoli, Domizia. "Il Miraggio Della Buona Architettura." Il Giornala, 29 October 2004, p. 31.
"Jesolo, East Coast." CasAmica, 11 December 2004.
"Jesolo come Miami?" Casa Vogue, December 2004.
Spinelli, Carlo. "In laguna puntano in alto." Il Giornale Dell' Architettura, December 2004.
Gu, Gi. "Hobag monopolizza Jesolo." Il Sole 24 Ore, 13 November 2004.
Scarane, Di Simonetta. "Meier cambia volto a Jesolo." Italia Oggi, 29 October 2004.
Zambon, Martina. "Torri in vetro di venti piani Jesolo sara come Malibu." Corriere del Veneto, 28 October 2004.
Vallora, Marco. "Tre torri di Meier per rilanciare Jesolo." La Stampa, 28 October 2004.
Zangrando, Alessandro. "Meier: cosi il grattacielo cambiera il volto de Jesolo." Correier del Veneto, 11 September 2003, p. 12.
"Richard Meier & Partners, Jesolo Lido Village." L'Arca, April 2008, pp. 24–29.

Malibu Beach House
Reginato, James. "Water Shed." W Magazine, October 2006, pp. 328–333.
Hawthorne, Christopher. "A Wider Shade of Pale." Los Angeles Times, 23 July 2006, pp. 1, E44.

On Prospect Park
Aguilar, Andrea. "Las casas blancas." El Pais Semanal, 10 May 2009.
Seward, Aaron. "Rise of the Transparent Condo in New York City: The Unprivate House." The Architect's Newspaper, 20 June 2007, pp. 20–21.
Barry, Tina. "Good All Over." The Brooklyn Paper, 1 June 2007.
Dykstra, Katherine. "On the Market: The Hottest New Buildings Rising in N.Y.C." New York Post, 28 September 2006, p. 56.
Marshall, Alex. "New Apartments in NYC Have Something New: Architecture." Regional Plan Association, 22 September 2006.
Abelson, Max. "Richard Meier, Meet SDS Procida (and 'Intelligent Design')." New York Observer, 14 August 2006.
Kolben, Deborah. "Luxe Glass Tower Planned for Prospect Heights." The Daily News, 12 April 2005.
Kolben, Deborah. "Tower to Rise Near P'Park." The Brooklyn Paper, 21 August 2004, pp. 1, 14.
Aleksander, Irina. "At Graydon Carter's Party, Swells Swill as Stocks Slide." New York Observer, 16 September 2008.
Aleksander, Irina. "Richard Meier on Real Estate Market: 'Fortunately I'm Just the Architect.'" New York Observer, 27 October 2008.
Rice, Andrew. "Betting On Star Power." Key: New York Times Real Estate Magazine, Spring 2007, pp. 84–89.
Seward, Aaron. "Rise of the Transparent Condo in New York City: The Unprivate House." The Architect's Newspaper, 20 June 2007, pp. 20–21.

Richard Meier & Partners Model Museum
Pogrebin, Robin. "Room With a View of an Architect's Retired Ideas." New York Times, 26 April 2007, pp. E1, E7.

Rothschild Tower
Lewis, Christina. "Tel Aviv Condos to Sell for $15 Million Each." Wall Street Journal, 17 October 2008.
Brass, Kevin. "Famed Architect Richard Meier on His New Project in Tel Aviv and the State of Tower Design." International Herald Tribune, 28 May 2008.
Bookatz, Karen. "If You Build It: Tel Aviv's 'Meier on Rothschild.'" Heeb, Summer 2008, pp. 30–31.

Saint-Denis Office Development
"Aboe Hac." 100% Office, Febuary 2007, p. 25.

San Jose City Hall
Conrad, Katherine. "City Hall Wins Design Award." San Jose Mercury News, 31 October 2006, pp. C1, C3.
Fukasaku, Rumi. "Richard Meier." Casa Brutus, December 2005, p. 25.
Foo, Rodney. "A Grand Opening." San Jose Mercury News, 14 October 2005, pp. 1A, 17A.
Jung, Monika. "A Skyline Redesigned." San Jose Mercury News, 13 October 2005, pp. 1, 11.
Hess, Alan. "North First Street Won't Work As a High-rise Downtown." San Jose Mercury News, 8 January 2006.
Berton, Brad. "History of Project Shows Many People Clearing Many Hurdles." Silcon Valley/San Jose Business Journal, 18 March 2005.
Berton, Brad. "City Hall Project Nears Finish Line." Silcon Valley/San Jose Business Journal, 18 March 2005.
Hess, Alan. "Vibrant Vision." San Jose Mercury News, 18 July 2005.
"Building San Jose's New City Hall." San Jose Mercury News, 18 July 2005.
Ostrom, Mary Anne. "The Architect: With Light-infused Design, Meier Sought Metaphor for Government Transparency." San Jose Mercury News, 18 July 2005.
Foo, Rodney. "City Hall Begins to Take Shape." San Jose Mercury News, 12 October 2003, pp. 1B, 4B.

Weill Hall, Cornell Unversity
Aloi, Daniel. "Architect Meier Speaks about Creating a Space Where People Are Happy Working Together." Cornell's Quarterly Chronicle, Winter 2009, pp. 16–17.
Ramanujan, Krishna. "At Weill Hall Dedication, Faculty Panel Ponders Life Science Feats." Cornell Chronicle, 24 October 2008, p. 3.
Campbell, Jennifer. "Weill Hall and Institute Dedicated in Celebration of "An Icon for Our Future."" Cornell Chronicle, 14 October 2008.
Lawyer, Liz. "Weill, Trustees Visit New Science Hub." Ithaca Journal, 17 October 2008, pp. 1A, 4A.
Ramanujan, Krishna. "Weill Hall Opens to Explore Life Science Frontier." Cornell Chronicle, 10 October 2008, pp. 1, 10.
Dang, Ming. "Life Sciences Building on Schedule." Cornell Chronicle, 27 September 2006.
"Ice Breaker: Ceremony Kicks Off Construction." Cornell Alumni Magazine, May/June 2005, p. 9.
Morisy, Michael. "C.U. Breaks Virtual Ground." Cornell Daily Sun, 14 March 2005.
Margolis, Michael. "Board Approves New Life Sciences Building." Cornell Daily Sun, 23 February 2005.
Brand, David. "Life Science Building Could Get Its Start Early Next Year, Architect Says." Cornell Chronicle, 20 May 2004, pp. 1–2.
Friedlander, Blaine. "Meier Unveils Proposed Design for Life Science Technology." Cornell Chronicle, 15 January 2004.
Ulrich, Claire. "New at the New Life Sciences Initiative." Cornell Communique, Fall 2003, pp. 18–19.
Brand, David. "CU Life Science Building Now Scheduled for Completion in Less than Four Years." Cornell Chronicle, 5 June 2003, p. 3.
Shea, Jennifer. "Life Span." Cornell Communique, Fall 2002, pp. 4–11.
"Trustees Approve Alumni Field as Site for New Life Science Technology Building." Cornell Chronicle, 1 February 2002.
Brand, David. "New Building Promises to Be 'Magnet' Providing Connectivity and Education." Cornell Chronicle, 1 February 2002.
"Trustees Approve Alumni Field as Site for New Life Science Technology Building." Cornell Chronicle, 31 January 2002.

合作者
Collaborators

New York Office

Tetsuhito Abe
Luca Aliverti
Daisy Ames
Kevin Baker
Reja Bakhshandegi
Carlo Balestri
Jeff Barajas
John Bassett
David Bench
Jonathan Bell
Remy Bertin
Ron Broadhurst
Cecile Brouillaud
Mary Lou Bunn
Marcus Carter
Carol Chang
Peter Choi
Gerard Chong
John Clappi
Joshua Coleman
Cedric Cornu
Maria Cumella
Adam Cwerner
Thibaut Degryse
Bryce de Reynier
Gong Dong
Alfonso D'Onofrio
Victor Druga
Nina Edelman
Jerome Engelking
Gil Even-Tsur
Simone Ferracin
Monica Franklin
Laura Galvanek
Clarisa Garcia-Fresco
Maria Gavieres
Dana Getman
Joanna Gontowska
Adam Greene
Kevin Hamlett
Eva Held
Clive Henry
Aron Himmelfarb
Yun Hsueh
Julianne Icart
N. Scott Johnson
Christopher Karlson

Bernhard Karpf
Robert Kim
Warren Kim
Lisetta Koe
Nikolas Koenig
Matthew Krajewski
Chung-Lun Kuo
Dilge Kutuoglu
Adrien Lambert
Thierry Landis
Chris Layda
Jason Lazarz
Dongkyu Lee
Kevin Lee
Vivian Lee
Robert Lewis
Grace Liao
Reynolds Logan
James Luhur
Roberto Mancinelli
Jose Gabriel McKinney
Ana Meier
Marianna Mello
Hyung Moon
Valentina Moretti
Emily Mottolese
Guillermo Murcia
Takumi Nakagawa
Adam Nicholson
Michael Norton
Sean O'Brien
Ringo Offerman
Matthias Opplinger
David Paz
Hans Put
Sarah Pyle
Robert Ramirez
Bettina Regensberger
David Robins
Amalia Rusconi-Clerici
Thomas Ryan
Stefan Scheiber-Loeis
Veronika Schmid
Tobias Schneberger
Michael Schneider
Judith Shade
Carl Shenton
Irina Sorrentino
James Stephenson

Bernhard Stocker
Anne Strüwing
Hyun-Young Sung
Yuri Suzuki
Justina Szal
Carlos Tan
Michael Taranto
Christopher Townsend
Michael Trudeau
Christian Tschoeke
Wen-Yu Tu
Tsung-Ming Tung
Hung-Kuo Wu
Dukho Yeon
Aaron Vaden-Youmans
Isabel Van Haute
Logan Werschky
Janet Yoder
Hyunjoon Yoo
San You

Los Angeles Office

Yasaman Barmaki
Therese Bennett
Stephen Billings
Karen Bragg
Jacquelyn Cacan
David Chang
Hyuk Steven Chung
James R. Crawford
Rhonna Del Rio
Lori East
Tom Farrell
Rozan Gacasan
Shekar Ganti
Farshid Gazor
Stefan Gould
John Gralewski
Michael Gruber
Suh-Yung Hahn
Vanessa Hardy
Ryan Indovina
Richard Kent Irving
Seung Jo
Yunghee Kim
Philip Koss
Elizabeth Kountzman
Danish Kurani

Marlene Kwee
Hyunseok Lee
Joongkee Lee
Kimberly Lenz
Grace Liao
Zhiwei Liao
Markita Littlejohn
Peggy Liu
Stuart Magruder
Chester Nielsen
Aryan Omar
Michael J. Palladino
Neda Rouhipour
Timothy Shea
Kristen Smith
Mark Sparrowhawk
Richard Stoner
Yuri Suzuki
Jonathan Swift
Nathan Urban
Matthew Uselman
Elise Wall
Alex Wuo

顾问
Consultants

ANC, Inc.
Aaron Pine,
Construction Specifications, Inc.
BR+A Mechanical Engineers
Buro Happold
CUBOID
Curtain Wall Design
and Consulting, Inc.
Davis Langdon &
Seah International
DEGW
dbox
Draheim Ingenieure
Ehrensberger & Oertz Architekten
Englekirk and Sabol
Consulting Structural, Inc.
Entek Engineering, Llt.
Erco Light Scout
Field Operations
Fisher Marantz Stone, Inc.
Follis Design
Gardiner & Theobald
Gennaro Guala
Gordon H. Smith Corporation
HKA Elevator Consulting, Inc.
IBE Consulting Engineers, Inc.
Jacobs Consultancy
JENIK
John Eisler Architects
Kimley-Horn & Associates, Inc.
Kirkegaard Associates
L'Observatoire International
Mas et Roux Architectes
Newson Brown Acoustics LLC
Olin Partnership
Ove Arup & Partners Ltd.
Pamela Burton & Company
Landscape Architecture
Paula Sekles
Peter Kruse Architekt
Piscatello Design Centre
Psomas Inc.
R.A. Heintges & Associates
R. Hough Architectural Concrete
RKW Architekten
Roofing & Waterproofing Forensics, Inc.
Schirmer Engineering
Serving, s.r.l

Shen, Milsom & Wilke, Inc.
Sherman Kung & Associates
Shimahara Illustration
Spojprojekt Praha, a.s.
Spurlock Poirer Landscape Architects
Steven Winter Associates, Inc.
Design Studio HoráK
Studio Commerciale Sonzogni
Studio Verdina
Syska Hennessy Group
Van Deusen & Associates
Viridian Energy & Environmental, LLC
WTM Ingenieurbüro

图片注解
Illustration Credits

Photographs and renderings of buildings are by:

© Richard Bryant, p. 406

© Jack Coyier Photography, p. 402

© dbox, pp. 58–59, 63–65, 114–115, 282–285, 290–291, 302–303, 310–315, 350–351, 358–361, 366–367, 396, 406, 412

© ECM, p. 252

© Face 2 Face Studio, pp. 162–163, 168–169, 400

© Klaus Frahm, pp. 92–93, 96–97, 210–211, 216–217, 410

© Scott Frances, pp. 228–229, 236–245, 342–343, 346–349, 408, 412, 414–415, 417–421, 408

© Scott Frances/ESTO, pp. 19–20, 22–25, 38–39, 44–57, 178–179, 184–186, 188–189, 398, 400

© Scott Frances, courtesy of Architectural Digest Copyright 2009 Conde Nast Publications Inc., 98–99, 102–105

© Shimahara Illustration, pp. 292–293, 300–301, 324–325, 330–333, 396, 398

© Studio AMD, pp. 118–122, 124–129

© Tim Griffith, pp. 26–27, 32–37, 187, 190–191, 196–197, 400, 402, 410

© Roland Halbe, pp. 142–143, 150–161, 170–177, 198–199, 206–209, 212–215, 218–219, 224–227, 246–247, 253–257, 266–267, 272–281, 396, 404, 408

© Phillip Kudelka, p, 187

© Wolfgang Ludes, p. 408

© Richard Meier & Partners Architects, pp. 66–67, 72–75, 79–80, 83–84, 86–87, 106–107, 112–113, 130–131, 136–139, 334–335, 340–341, 367–369, 374–377, 380–385, 398, 400, 402, 404, 412

© Jock Pottle, pp. 316–317, 322–323, 396

© Tomas Riehle, pp. 258–259, 264–265, 398

© Juan Pablo Vergara, p. 136

All drawings by
Richard Meier & Partners Architects.

All sketches by Richard Meier.

Portrait of Richard Meier on page 388 is © Mark Seliger.